MIXED DESIGN OF INTEGRATED CIRCUITS AND SYSTEMS

AF850501

THE KLUWER INTERNATIONAL SERIES IN ENGINEERING AND COMPUTER SCIENCE

MIXED DESIGN OF INTEGRATED CIRCUITS AND SYSTEMS

edited by

Andrzej Napieralski
Zygmunt Ciota
Technical University of Łódź
Łódź, Poland

Augustin Martinez
Laboratoire d'Analyse et d'Architecture des Systèmes
du Centre National de la Recherche Scientifique
Toulouse, France

Gilbert De Mey
University of Gent
Gent, Belgium

Joan Cabestany
Universitat Politecnica de Catalunya
Barcelona, Spain

KLUWER ACADEMIC PUBLISHERS
Boston / Dordrecht / London

Distributors for North America:
Kluwer Academic Publishers
101 Philip Drive
Assinippi Park
Norwell, Massachusetts 02061 USA

Distributors for all other countries:
Kluwer Academic Publishers Group
Distribution Centre
Post Office Box 322
3300 AH Dordrecht, THE NETHERLANDS

Library of Congress Cataloging-in-Publication Data

A C.I.P. Catalogue record for this book is available
from the Library of Congress.

Copyright © 1998 by Kluwer Academic Publishers

All rights reserved. No part of this publication may be reproduced, stored in a retrieval system or transmitted in any form or by any means, mechanical, photo-copying, recording, or otherwise, without the prior written permission of the publisher, Kluwer Academic Publishers, 101 Philip Drive, Assinippi Park, Norwell, Massachusetts 02061

Printed on acid-free paper.

CONTENTS

III. Microsystems and Neural Networks

IV. Design Methodologies

V. Advanced Trends in Microelectronics Education

Preface

Very fast advances in IC technologies have brought new challenges into the physical design of integrated systems. The emphasis on system performance, in lately developed applications, requires timing and power constraints to be considered at each stage of physical design. The size of ICs is decreasing continuously, and the density of power dissipated in the circuits is growing rapidly.

The first challenge is the Information Technology where new materials, devices, telecommunication and multimedia facilities are developed. The second one is the Biomedical Science and Biotechnology. The utilisation of bloodless surgery is possible now because of wide micro-sensors and micro-actuators application. Nowadays, the modern microsystems can be implanted directly into the human body and the medicine can be applied right in the proper time and place in the patient body. The low-power devices are being developed particularly for medical and space applications. This has created for designers in all scientific domains new possibilities which must be handed down to the future generations of designers.

In this spirit, we organised the Fourth International Workshop "MIXED DESIGN OF INTEGRATED CIRCUITS AND SYSTEMS" in order to provide an international forum for discussion and the exchange of information on education, teaching experiences, training and technology transfer in the area of microelectronics and microsystems.

The objective of this Conference was to discuss all the problems which have to be taken into account by future designers of modern integrated circuits and devices. Different aspects of the advanced electronic design, testing and manufacturing, which should be reflected in the high-level education, have been presented. Therefore the topics of this Conference have been very carefully chosen. The main topics of discussion were:

- Design methodologies, techniques, and practical examples
- Thermal problems in semiconductor devices and ICs
- Analog and digital filters and signal processing
- Advanced testing methods and reliability
- Neural networks
- Sensors, actuators and micromachines: modelling, simulation and design
- Power devices and Smart-Power modules
- New trends in microelectronics: RF, Low Power, Low Voltage
- Education of CAD of modern IC devices
- EUROPRACTICE activities

All these topics are very important for the general knowledge of students. Future engineers must be able to think not only about the common utilisation of the specific VLSI design tools, but they have to think about all their design environment as well. In modern electronics, testing problems, thermal problems, sensors and actuators design problems and all aspects of power integration design are of great importance.

The Conference covered the most interesting problems in the microelectronics design in a wide range of topics variety. The programme of the Conference included prepared invited presentations, oral presentations and poster sessions. The best presentations have been carefully chosen as the contents of this book. It is the reflection of issues and views debated at the Conference and the summary of technical assessments and results presented.

The Conference was organised by the programme committee, and all the papers presented at the Conference have been reviewed by at least two members of programme committee. The best papers of each session were chosen for this book by the session chairmen and then reviewed again by the following members of International Steering Committee:

Prof. A. Napieralski, Technical University of Łódź, Poland

Dr Z. Ciota, Technical University of Łódź, Poland

Prof. A. Martinez, Laboratoire d'Analyse et d'Architecture des Systèmes du Centre National de la Recherche Scientifique, France

Prof. G. De Mey, University of Gent, Belgium

Prof. J. Cabestany, Universitat Politecnica de Catalunya, Spain

Prof. A. De Vos, University of Gent, Belgium

Prof. R. Ubar, Tallinn Technical University, Estonia

Dr J. L. Noullet, Institut National des Sciences Appliquées de Toulouse, France

Prof. M. Glesner, Technische Hochschule, Darmstadt, Germany

Prof. W. Kuźmicz, Warsaw University of Technology, Poland

Prof. A. Handkiewicz, Poznań University of Technology, Poland

Dr J. M. Moreno, Universitat Politecnica de Catalunya, Spain

This book groups together the selected papers (only 25% of presented paper have been chosen) written by experts in technology, modelling, analogue and digital filters design, signal processing, neural network design, thermal design, design methodologies, electromagnetic compatibility, sensors and actuators, advanced testing methods and reliability, power devices and Smart-Power modules. The special emphasis has been laid on the educational aspects of all presented problems. The book is divided into the following 5 parts:

1. Analog Circuit Design
   (coordinated by **Prof. G. De Mey and Dr Z. Ciota**)
2. Power Devices & Thermal Aspects
   (coordinated by **Prof. A. Napieralski and Prof. G. De Mey**)
3. Microsystems & Neural Networks
   (coordinated by **Prof. A. Martinez and Prof. J. Cabestany**)
4. Design Methodologies
   (coordinated by **Prof. J. Cabestany and Dr J.L. Noullet**)
5. Advanced Trends in Microelectronics Education
   (coordinated by **Dr J.L. Noullet and Dr Z. Ciota**)

We hope that you will find this book interesting and it will be useful aid for all students and engineers involved in VLSI circuits design.

The Conference was sponsored by Poland Section IEEE - CAS Chapter, INCO-Copernicus SYTIC Project, Polish State Committee for Scientific Research, ESPRIT-BARMINT Program and Australasian Association for Engineering Education.

For all those who came together to share ideas, and for all the prospective readers, we hope that this publication will excite engineers and students to search new ideas.

Łódź, Poland	Andrzej NAPIERALSKI
Łódź, Poland	Zygmunt CIOTA
Toulouse, France	Augustin MARTINEZ
Gent, Belgium	Gilbert DE MEY
Barcelona, Spain	Joan CABESTANY

PART I

Analog Circuits Design

1

MULTICHANNEL LOW NOISE, LOW POWER ANALOGUE READOUT CHIP FOR SILICON STRIP DETECTORS

Wojciech Białas, Władysław Dąbrowski, Paweł Gryboś and Marek Idzik

Faculty of Physics and Nuclear Techniques
University of Mining and Metallurgy
al. Mickiewicza 30, 30-059 Kraków
POLAND

ABSTRACT

A prototype 16-channel readout chip for a X-ray detection system using silicon strip detectors is presented. The chip has been designed as a full custom ASIC for the AMS 1.2µm CMOS process. Single channel of the circuit consists of low noise, medium speed preamplifier, followed by a shaper and a discriminator. Such a system allows to use silicon strip detectors in the single photon counting mode. Key design issues and optimization of critical design parameters are discussed. Measurement results from the successfully manufactured prototype are presented and discussed.

1. INTRODUCTION

A silicon strip detector is an array of reverse biased junction diodes made on high resistivity silicon substrate. The diodes are shaped as strips with pitch from 20µm to 200µm according to the geometrical requirements of an experiment. A particle crossing the active detector region loses its energy and produces short current pulses on the strips in the vicinity of particle trajectory. When silicon strip detectors are used for X-ray detection in applications like X-ray imaging or X-ray diffractometry they work in the single photon counting mode. An effective use of these sensors requires a special fast, low noise readout electronics which essentially can be designed only as full custom VLSI ASICs. In this paper we present the design and the measurement results of a prototype multichannel chip optimized for readout of silicon strip detectors for a Roentgen diffractometer.

In the presented design a number of issues relevant for most analogue VLSI circuits, like low-noise, low-power and matching, is addressed carefully. A current signal generated in a strip of silicon detector by a single photon is amplified, shaped and applied to a comparator providing 1-bit yes/no information. For such a signal processing scheme a sufficient separation between the noise level and the signal level is required in order to provide a high detection efficiency and a limited noise count rate. For a system with a first order band-pass filter of time constant τ the noise count rate at the comparator output is given by Rice formula in [1].

$$f_n = \frac{1}{2} f_{n0} \exp\left(-\frac{V_t^2}{2V_n^2}\right) \tag{1}$$

where V_t is the comparator threshold, V_n is the RMS value of noise at the comparator input and f_{n0} is zero crossing frequency which equals to $0.29/\tau$ [2]. In order to keep the noise count rate at a negligible level, i.e. 10^{-3} count/sec or below, it is required for a typical detection system that the comparator threshold is set at least at a level of $3V_n$. On the other hand the comparator threshold should be low enough to provide full efficiency for real signal counts. Since the signal is smeared by noise then another margin of $3V_n$ for the threshold setting is needed. Then, taking into account some additional effects like fluctuation of charge generated in silicon detectors, charge division between neighbouring strips and matching of amplifier gain and comparator offset for a multichannel system, one arrives to a conclusion that a signal-to-noise ratio of 10 is required at minimum. In order to make a multichannel system practical one requires that a common discrimination threshold is applied for all channels. Therefore the matching of single channel parameters in a multichannel chip is an absolutely critical issue for such a system.

A typical energy of X-rays used in Roentgen diffractometry is 8 keV. A photon of this energy when absorbed in the depletion region of a silicon diode produces of about 2000 electron-hole pairs which are collected by the readout strips within the time period of about 20 ns [3, 4]. In an ideal situation one expects the signal at the amplifier output to be proportional to the charge collected in the detector. Thus for such a system we define the charge gain as the ratio (output voltage signal)/(input charge) and the equivalent noise charge (ENC) as the charge applied to the input in a form of a short δ-like current pulse which gives at the output a signal amplitude equal to the RMS value of noise V_n.

2. SINGLE CHANNEL CIRCUIT

The block diagram and full circuit implementation of a single channel is shown in Figure 1. It comprises four blocks: preamplifier, shaping amplifier, differential stage and discriminator.

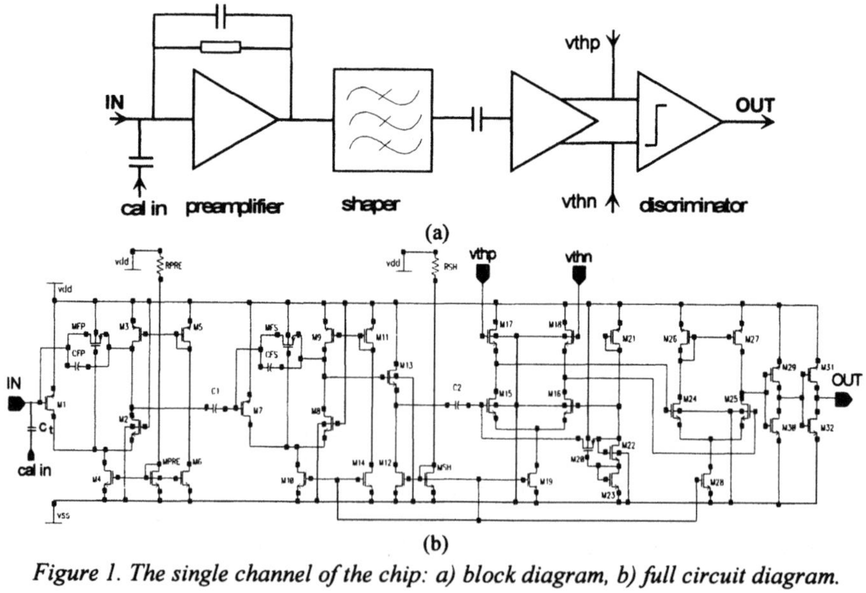

Figure 1. The single channel of the chip: a) block diagram, b) full circuit diagram.

The preamplifier stage is the charge sensitive amplifier which integrates the current input signal from a silicon strip detector into a voltage signal. The signal at the preamplifier output is a step function with exponential decay defined be the long time constant of the feedback loop. The voltage signal is applied to a band-pass filter which provides two functions: (i) shaping of a short pulse according to timing requirements and (ii) limiting of the bandwidth to a minimum corresponding to the signal spectral density in order to keep minimum noise bandwidth.

In order to optimize the noise performance of the chip, the preamplifier and the shaper is made in a single ended configuration. For such a system and a relatively short peaking time, when the 1/f noise is negligible, the equivalent noise charge ENC expressed in number of electrons is given as [5]:

$$ENC = \sqrt{\frac{F_v v_n^2 C_{in}^2}{T_p} + F_i i_n^2 T_p} \qquad (2)$$

where:

C_{in} is the total input capacitance including the gate capacitance of the input transistor, silicon detector capacitance and any stray capacitance introduced by connection between the silicon strip detector and the readout chip,

v_n^2 is the spectral density of the equivalent input voltage noise, dominated by the equivalent input noise of the input transistor,

i_n^2 is the spectral density of the equivalent input current noise dominated by the thermal noise of the feedback resistor and the shot noise of the detector leakage current,

T_p is the peaking time, i.e. the time at which the signal at the filter output reaches the maximum; for a simple CR-RC band-pass filter peaking time T_p is equal to the time constant of the filter $\tau = RC$

F_v and F_i are constants dependent on the filter type.

Two immediate observations based on the formula (2) are important for optimization of the front-end circuit, namely: (i) contribution of the input voltage noise to ENC is proportional to the total input capacitance and inversely proportional to the square root of the peaking time and (ii) contribution of the input current noise is independent of the input capacitance and proportional to the square root of the peaking time. One can now optimize the front-end system for various requirements and constrains given by a particular application. If for example there is no constraint on the peaking time, like for any low rate experiment, one can find an optimum peaking time for given other parameters, voltage and current noise spectral densities and detector capacitance. However, in many applications high counting rate capability is a serious requirement which has to taken into account as a limitation for the peaking time. Another parameter driven strongly by applications is the detector capacitance. Thus, in most cases in order to minimize ENC one has to focus on the optimization of the input transistor.

Equivalent input voltage noise of a MOSFET is inversely proportional to the transconductance g_m. In strong inversion g_m is proportional to $\sqrt{(W/L)I_d}$, where W, L, I_d are the width, length and drain current of the input transistor respectively. To minimize the voltage noise W/L and I_d should be as large as possible, however, in order to find an optimum W/L one has to take into account some other effects. First of all a minimum L allowed by a given technology should not be used in order to avoid an excess noise due to short channel effects [6]. For a given gate length L the input capacitance of the transistor will increase proportionally to the gate width W and will contribute to the total input capacitance in the formula (2).

For our design we have chosen peaking time of 550ns based on requirements concerning the maximum X-ray intensity per single strip. Taking into account the foreseen detector capacitance of about 2 pF per strip and all other aspects discussed above we have chosen a PMOS of W/L=1500μm/1.5μm as an input device. For the chosen peaking time one can expect some extra contribution to ENC due to 1/f noise which is neglected in formula (2). Since it is well known that 1/f noise is usually lower for PMOS transistors compared to NMOS transistors we have used a PMOS one although this solution has a small drawback due to a higher current required for a given transconductance.

The preamplifier and the shaper are based on a folded cascode configuration which provides good open loop gain and bandwidth as well as good power supply rejection ratio. The last aspect is quite important since due to noise reason the preamplifier has to be design in a single ended configuration. The currents in the preampiifier stage and in the shaper are controlled by the external resistors RPRE and RSH (see Figure 1) from several μA to several hundred μA. This solution gives a possibility to control the gain, the peaking time and the power consumption and to optimize in some range the settings for different detector capacitances.

Another drawback of the single ended configurations of the preamplifier and the shaper results in significant channel-to-channel and chip-to-chip offset variation Therefore AC coupling is implemented between these two stages as well as between the shaper and the comparator. Furthermore the rest of the circuit, after the shaper, is designed in a fully differential mode. The differential pair after the shaper has a gain close to 1 and is basically used for providing a differential threshold to the comparator. The differential scheme for setting the comparator threshold allows to use the circuit either for positive or for negative input signal polarity. Providing the gain in the preamplifier and shaper are high enough the chosen solution allows us to apply a common threshold for all 16 channels in the chip and keep the comparator offset variation negligible compared to noise.

3. LAYOUT

The circuit is laid out as 16-channel chip with 90 μm pitch. The length of the chip including input and output bond pads is equal 2.5 mm. All bond pads for the bias and control lines are placed on both sides of the chip. Along each channel a number of small probe pads has been placed which provide a capability of probing the signal after each stage of the circuit. In front of each preamplifier a test capacitor of 50 fF is placed. The test signals are distributed from two calibration pads to every second channel from each. This scheme provides a possibility for testing the cross-talk between channels.

4. TEST RESULTS

The chip has been manufactured successfully in the AMS 1.2μm process. The basic functionality of a single channel is illustrated in the Figure 2. The measurements were performed using the internal calibration circuitry. Applying a voltage step signal V_{in} to the calibration capacitor C_t which is connected to the input of every channel is equivalent to an injection of the charge $Q_{in} = C_t V_{in}$. Figure 2 shows the signal wave forms measured at the output of the shaper and at the output of the discriminator for an input signal of 1600 electrons. The upper plot shows results of measurements with a digital scope in the averaging mode, while the lower plot shows the result of a single shot measurements accumulated over some time period. The intensity of the shaper output wave form gives a rough estimate of signal-to-noise ratio of about 20. Precise measurements of noise give a typical ENC value of about 50 electrons RMS for zero detector capacitance. For the nominal bias currents the power consumption is below 3 mW/channel.

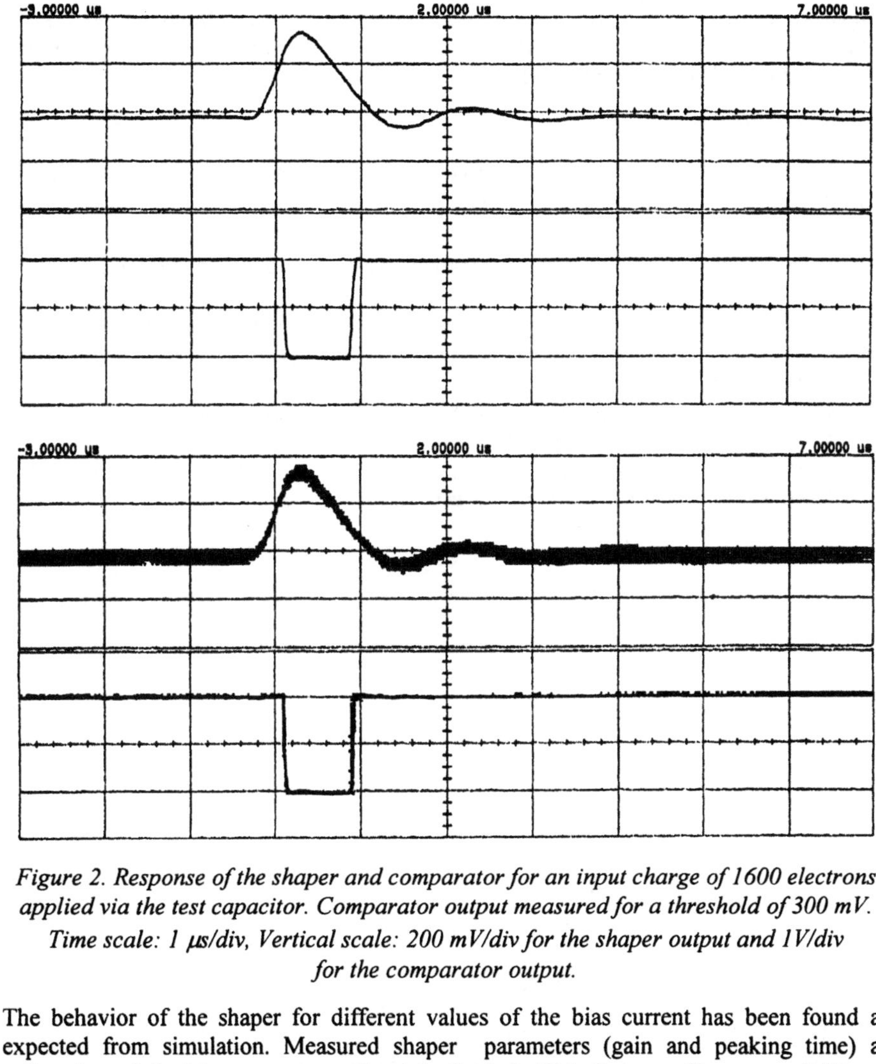

Figure 2. Response of the shaper and comparator for an input charge of 1600 electrons applied via the test capacitor. Comparator output measured for a threshold of 300 mV. Time scale: 1 μs/div, Vertical scale: 200 mV/div for the shaper output and 1 V/div for the comparator output.

The behavior of the shaper for different values of the bias current has been found as expected from simulation. Measured shaper parameters (gain and peaking time) as a function of the bias current are listed in Table 1. This feature offers a possibility to minimize the noise for different detector capacitances by choosing a proper peaking time.

Table 1. Measured charge sensitivity and peaking time as a function of the shaper current.

Current in the shaper [μA]	**Charge sensitivity [mV/fC]**	**Peaking time [ns]**
40	650	800
60	580	550
80	400	400

5. CONCLUSIONS

A low noise, low power ASIC for readout of silicon strip detectors has been designed and manufactured successfully. The experimental results obtained form the prototype are in good agreement with the designed parameters. It has been shown that using the concept employed in the presented circuit one can design a very low noise front-end circuitry for silicon strip detectors allowing for efficient detection of X-ray photons of energy as low as 8 keV in the single photon counting mode.

REFERENCES

[1] S. O. Rice, *Mathematical Analysis of Random Noise*, Bell System Tech. J., 24 (1945), p.55

[2] H. Spieler, *Power Requirements for Frontend Electronics in the Silicon Tracker,* SCIPP Report 91/28, September, 1991.

[3] J. Kemmer and G. Lutz, *New detector concepts,* Nucl. Instr. and Meth. Vol. A 253, (1987), p. 365.

[4] W. Dabrowski, P. Grybos and M. Idzik, *Study of spatial resolution and efficiency of silicon strip detectors with different readout schemes,* Nucl. Instr. and Meth. Vol . A 356 (1995), p. 241.

[5] E. Gatti and P.F. Manfredi, *Processing the signal from solid-state detectors in elementary particle physics,* La Revista del Nuovo Cimento, Vol. 9, No.1, 1986.

[6] W. Dabrowski et al. *Noise measurements on radiation hardned CMOS transistor,* SCIPP Report 91/94, University of California, Santa Cruz, 1991.

2

DESIGN METHODOLOGY FOR CURRENT CONVEYOR BASED CONTINUOUS-TIME FIELD-PROGRAMMABLE ANALOG ARRAY

Richard Grisel, Christophe Premont, Nacer Abouchi and Jean-Pierre Chante

CPE Lyon, LISA CNRS EP 0092
43 Bd. du 11 Nov. 1918, BP 2077 Villeurbqnne
FRANCE

ABSTRACT

A design methodology for continuous time Field-Programmable Analog Array (FPAA) is presented. After introducing the key features of FPAA and dealing with design issues related to continuous-time applications, we present the elementary cell of the proposed FPAA. This cell is based on current conveyors and designed for high-frequency applications. Two examples are presented: a high-frequency amplifier and a high-frequency multiplier.

1. INTRODUCTION

One defines a Field-Programmable Analog Array (FPAA) as an integrated circuit which allows, by the mean of flexible programmability facilities, to implement analog functions. If we consider the recent market increase of programmable devices such as PALs, EPLDs, FPGAs for the digital counterpart, it is obvious to say that, for real-time signal processing applications, FPAA design should provide electronics designers with a very efficient and powerful tool.

A FPAA consists of programmable analogue elementary cells which can be interconnected by the mean of programmable interconnections. The reconfigurable cell has to perform a set of functions, in order to provide the flexibility, with good electrical performance.

The set of functions is classically defined as: amplifier, comparator, multiplier, voltage-controlled oscillator.

Different previous approaches are operational amplifier-based and operate at limited bandwidth (100kHz) and have limited linearity due to the use of MOSFET based switches. The proposed approach tries to cope with these two major hints, by the use of a current conveyor based elementary cell for the reconfigurable analogue block, allowing high-frequency continuous-time applications, and being programmable as a pass or no-pass switch by tuning its bias sources.

2. DESIGN ISSUES FOR FIELD-PROGRAMMABLE ANALOG ARRAY

Design of analog applications requires attention to several parameters like noise level, distortion, dynamic range, etc. The two major key features for the design of an efficient and useful analog array are the programming of the elementary functions and the reconfigurable topology. The elementary cell function is set by changing some parameters like the value of programmable transconductances and capacitors. The programmed cell has to perform its function with good electrical performance offering wide parameter range for a flexible analog array. Two different approaches, the continuous-time or the switched approach, can be considered. These two techniques do not offer the same trade-offs between performance and parameter range for the programming.

The switched approach consists of switched-capacitors or switched-current which are digitally controlled and provide a wide parameter range but with frequencies limited to few hundred kilohertz because of the Shannon theorem [1,2,3].

The continuous-time approach offers a lower parameter range compared to the switched approach, but with particular design techniques, like the use of current conveyor for example, some very efficient analog blocks can be developed for the analog array. In the previous works [4,5] using the continuous-time approach for the design of field-programmable analog array, the performance of the circuit were limited by the use of both Op-Amp based design and analog switches, preventing high-frequency operation of the circuits.

A new methodology for the elementary analog cell design is introduced, and a specific approach for the interconnection of the cells without the use of switches in the signal path is addressed. Several properties of current conveyors are used to achieve both a high-frequency operation of the elementary cells and a local interconnection scheme. The configuration of the circuit, the programming of the functions and of the topology require the used of bit registers and control voltages. Some registers are converted with a digital to analog converter to provide control voltages to set the values of the programmable transconductances and capacitors of the array [6-7]. The control voltages are locally stored, near the element to program, on a capacitor which needs to be periodically refreshed. New configuration techniques have been recently addressed, as for example the use of EEPROM non-volatile analog memory. This technique has been thoroughly studied in [8]. The present paper focuses on the design of the analog cell, and the configuration of the analog array and its architecture are not addressed.

3. ELEMENTARY ANALOG CELL

The analog elementary cell of the array, presented in Figure 1, consists of a four MOSFET transconductor and two conveyor-based I-V converters. The four MOSFET transconductor [9-10] is a highly linear differential, programmable transconductor with response given by:

$$\left(I_{Xp} - I_{Xn}\right) = \mu C_{OX} \frac{W}{L} \left(V_{C1} - V_{C2}\right)\left(V_{Ep} - V_{En}\right) \tag{1}$$

Parameters μ and C_{OX} are respectively the average carrier mobility in the channel and the gate oxide capacitance per unit area. L and W are respectively the length and the width of the transistor. Relation (1) is true if the conditions $V_1, V_2 < \min\left[V_{C1} - V_T, V_{C2} - V_T\right]$ are verified. (V_t is the threshold voltage for a MOS transistor)

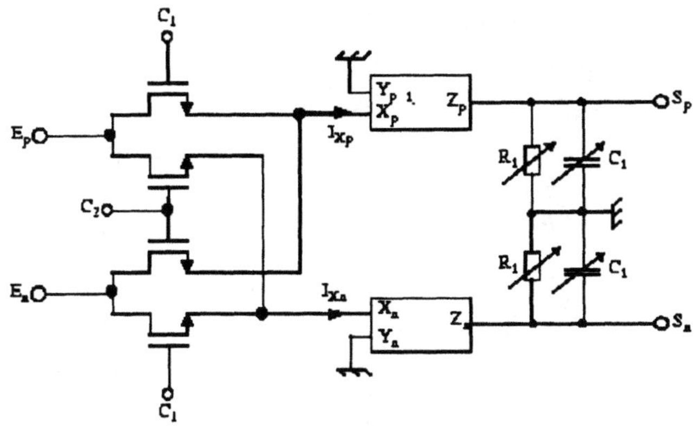

Figure 1. Elementary Analog Cell.

A current conveyor (CC) [11] is a three terminal device which operates in such a way that if a voltage is applied to the input terminal Y, an equal potential will appear on the input terminal X. In a similar fashion, an input current forced into terminal X will result in an equal current flowing into terminal Z.

The two CCs of Figure 1 perform the I-V conversion. Virtual grounds at nodes X have to be created in order to make the four MOSFET transconductor to behave linearly. The currents IXp and IXn flowing into the X nodes produce the voltages V_{Sp} and V_{Sn}. Using equation (1) the following relation is derived:

$$\frac{V_{Sp} - V_{Sn}}{V_{Ep} - V_{En}} = \frac{R_1}{1 + sR_1C_1} \mu C_{OX} \left(V_{C1} - V_{C2} \right) \qquad (2)$$

The difference (V_{c1}-V_{c2}) is used to perform a four-quadrant analog multiplier and to control the polarity of the output signal. The load resistor and capacitor are used to perform either an amplifier or an integrator or a low-pass first order filter.

The programmable resistor R_1 is a CMOS resistor with transistors operating in ohmic region [12], in order to have both good linearity and parameter range. The programmable capacitor is based on a capacitive multiplier thoroughly described in [13].

The design of an analog processing application is performed by cascading two CCs. The output port Z of the first CC is connected to the input port Y of the second CC. If several CCs port Z are connected to the same node, then the output currents are added and converted in voltage with a resistor before going to the next stage.

As explained in the next section, a CC is biased by two current sources, the positive and the negative one. A CC can perform interconnection between cells, simulating a pass switch or a non-pass switch by turning on or off respectively its two bias current sources [14]. The CC's equivalent input impedance is modified with the polarisation.

4. HIGH-FREQUENCY CONSIDERATIONS

In the presented analog elementary cell the two current conveyors are used to transfer the current from the four MOSFET transconductor to the load. The frequency limitation is due to the current transfer Iz/Ix. The schematic of a current conveyor, presented in Figure 2, shows that the current transfer between terminal X and Z is performed by two current mirrors M6-M8 and M7-M9. The voltage at node Y is copied to node X using the current mirrors M1-M2 and M3-M4. The transistors M10, M11, M12 and M13 are the bias current sources controlled by two voltages V_p and V_n.

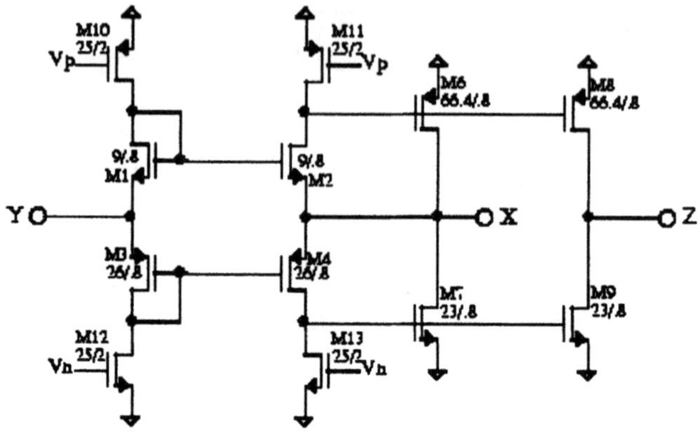

Figure 2. Schematic of the current conveyor

It can be shown that for the elementary current mirror of Figure 3, the approximate first order transfer functions is:

$$I_{OUT} = gV_{OUT} = -\frac{I_{IN}gm_2}{gm_1}\frac{1-s\frac{Cgd2}{gm_2}}{1+s\left(\frac{C+Cgd2}{g}+\frac{Cgs1+Cgd2}{gm_1}+\frac{gm_2Cgd2}{gm_1g}\right)}$$

$$\text{with } g=\frac{1}{R_L}+gds_2 \text{ and } C=Cdb_2+C_L$$

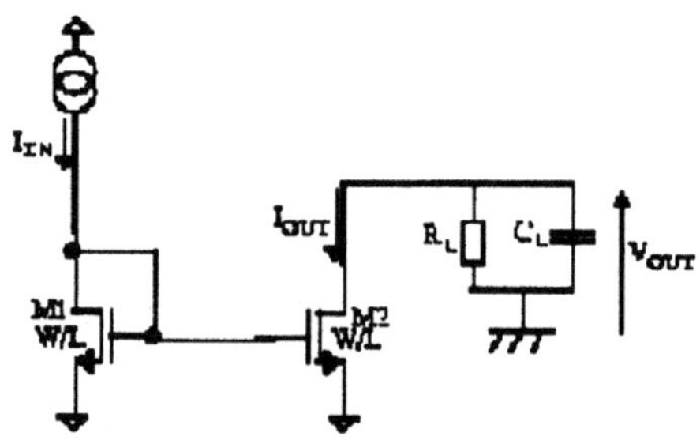

Figure 3.A simple current mirror.

gm_1, gm_2 are the transconductance of transistor M1 and M2 respectively. C_{gd2} and C_{db2} are the gate-drain capacitance and drain-bulk capacitance of transistor M2 respectively. C_{gs1} is the gate-source capacitance of transistor M1.

The main limitation are due to the C_{gd2} capacitor and to the load impedance of the current mirror. If low R_l and C_l value are used, referring to Figure 1, and reasonable W/L ratios are used to limit the C_{gd2} value, a high-frequency operation can be achieved. The current conveyor has been implemented using AMS 0.8um CMOS technology previously used in [15] and chosen for its stability as far as process parameters are concerned. HSpice simulations have been carried out and a bandwidth greater then 100MHz for the current transfer is achieved using the W/L ratios given in Figure 2.

5. RESULTS

The frequency response of the elementary cell (see Figure 4) used as an amplifier with a 20mV input signal for a voltage gain of 10 achieves 10Mhz bandwidth (-3dB cut-off frequency).

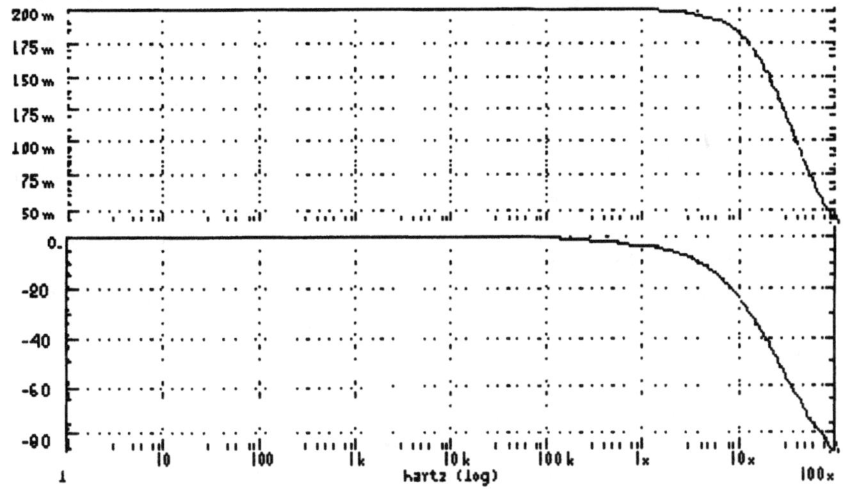

Output voltage: magnitude (upper curve), phase (lower curve)

Figure 4. C_{CS} based amplifier.

A frequency doubler can be performed using the four MOSFET transconductor as an analog multiplier and two input signals with the same frequency. Figure 5 shows that the circuit is able to perform well as a frequency doubler at 80MHz, for two 40MHz input signals.

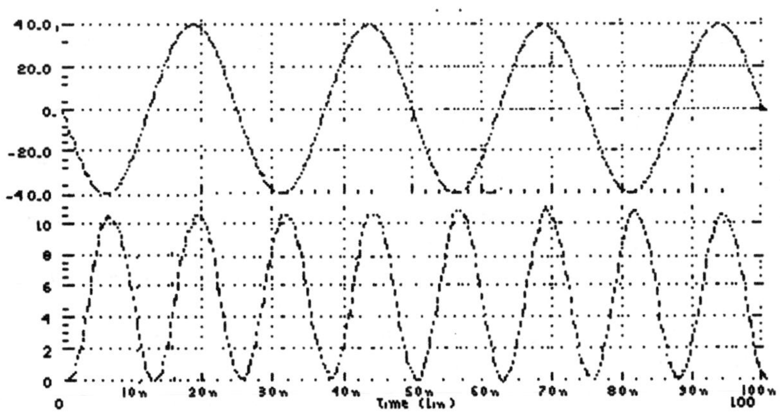

Input signal (upper curve), output signal (lower curve)

Figure 5: C_{CS} based frequency doubler.

6. CONCLUSION

A new approach for designing analog elementary cell for field-programmable analog array has been described. Its major improvement, compared to the previous works, is the use of current conveyors in order to achieve a wide range of analog functions operating at high frequency (80 MHz). The key performance feature of the proposed approach is the current-mode processing which seems to provide attractive solution for wide bandwidth capability.

ACKNOWLEDGEMENTS

The authors wish to thank the Scientific and Technology Cooperation Service of Ottawa French embassy in Canada for supporting part of this work.

REFERENCES

[1] A. Bratt and I. Macbeth, *Design and Implementation of a Field Programmable Analogue Array*, FPGA'96, pp. 88-93, 1996.

[2] H.W. Klein, *The EPAC Architecture: An Expert Cell Approach to Field Programmable Analog Array*, FPGA'96, pp. 94-98, 1996.

[3] S.T. Chang, B.R. Hayes-Gill and C.J. Paul, *Multi-function Block for a Switched Current Field Programmable Analog Array*, MWSCAS'96, Ames (Iowa), 1996.

[4] K.F.E. Lee and P.G. Gulak, *A CMOS Field-Programmable Analog Array*, ISSCC Digest of technical papers, pp. 186-188, Feb. 1991.

[5] K.F.E. Lee, *Field-Programmable Analog Arrays Based on MOS Transconductors*, Ph.D. Thesis, University of Tonronto, 1995.

[6] K.W. Current and M.E. Hurlston, *A Bi-Directionnal Current-Mode CMOS Multiple Valued Logic Memory Circuit*, Proc. of the Inter. Symposium on Multiple Valued Logic, pp. 196-202, May 1991.

[7] B. Hochet, V. Peiris, S. Abdo and M.J. Declercq, *Implementation of a Learning Kohonen Neuron Based on a New Multilevel Storage Technique*, IEEE Journal of Solid State Circuits, Vol. 26, No. 3, pp. 262-267, Mar. 1991.

[8] A. Thomsen and M.A. Brooke, *Low Control Voltage Programming of Floating Gate MOSFETs and Applications*, IEEE Trans. on Circuits and Systems, Vol. 41, No. 6, pp. 443-451, June 1994.

[9] Z. Czarnul, *Novel MOS Resistive Circuit for synthesis of Fully Integrated Continuous-time Filters*, IEEE Trans. on Circuits and Systems, Vol. 33, No. 7, pp. 718-721, July 1986.

[10] S.T. Dupuie and M. Ismail, *High Frequency CMOS transconductors*, in Analog IC Design: The Current-Mode approach, Peter Peregrinus Ltd., 1990.

[11] A.S. Sedra, C.W. Roberts and F. Gohh, *The Current-Conveyors: History, progress and new results*, IEE Proc., Vol. 137, Pt. G, No. 2, April 1990.

[12] J. Silva-Martinez, M. Steyaert and W. Sansen, *High-Performance CMOS Continuous-time filters*, Kluwer Academic Publishers, 1993.

[13] C. Premont, R. Grisel, N. Abouchi and J. P. Chante, *A Current Conveyor based Capacitive Multiplier* , MIXDES'97, Poznan, Poland, pp. 81-84, June 1997.

[14] C. Premont, N. Abouchi, R. Grisel and J.P. Chante, *A Current Conveyor based High-Frequency Analog Switch*, IEEE Trans. on Circuits and Systems, to be published in Dec. 1997.

[15] F. Bergouignan, N. Abouchi, R. Grisel, G. Caille and J. Caranana, *Designs of a Logarithmic and Exponential Amplifier Using Current Conveyor,* ICECS'96, Rodos, Greece, pp. 61-62, Oct. 1996.

3

RF AMP IC FOR OPTICAL DISC PLAYER

[1]Chun-Sup Kim, [1]Gea-Ok Cho,
[1]Yong-Hwan Kim and [2]Bang-Sup Song

[1]ASIC Center, Corporate Technical Operations
Samsung Electronics CO., Ltd
KOREA

[2]University of Illinois,
Urbana, IL
USA

ABSTRACT

This paper describes RF amp integrated circuit for the read channel of optical disc players that include CD and DVD demanding the equalizer and the decision block. A RF equalizer is composed of boosting filter to slim the data pulse and 9th order Bessel filter to attenuate high frequency noise. Boosting filter emphasize 6dB at 6MHz and maintains the linear phase within 1% by pole zero cancellation method. An important advantage is tuneability of the boosting and frequency characteristics by controlling the transconductance, therefore, it is enough to cover the tolerance of fabrication. The high frequency noise due to equalization is attenuated by 9th order Bessel filter which has the slope of over -42dB/oct. Measured jitter keeps within 4.5n. The decision block includes one bit analog to digital converter and the low pass filters to detect asymmetry and asymmetric amp. As the analog and the digital signals are mixed on a chip, the main ground and the NMOS bulk ground, DGND and AGND respectively, are separated. Also, it is laid out very carefully to minimize the digital noise in analog block. Implemented in a 0.8um CMOS n-well technology, the RF equalizer and the decision block chip occupy 3000um x 1000um. Power consumption from a single 5V is around 300mW.

1. INTRODUCTION

Recently the optical disc has become a choice of high-density storage medium for CD-ROM or digital video applications. In addition to its high capacity, the high-speed read capability has also become a key requirement for the success of the optical medium. The reading speed of the compressed MPEG-encoded data stored in optical discs, for example, is vital for their use. In the new DVD (Digital Versatile Disc) format, the storage capacity is 4.7GB and the data spectrum covers the frequency range of 5 to 6MHz. Compared to the existing CD format of 720kHz, the DVD needs higher speed electronics to compensate for high-frequency loss.

This paper presents the RF amp chip for high-speed optical discs and DVD players. The RF equalizer is integrated on one chip with decision block. It is implements by a Bessel filter with two symmetric zeros using a tuneable gm-C topology. The 9th order filter is chosen to effectively suppress high-frequency noise and to minimize the jitter. The decision block is made up of the one bit analog to digital converter and asymmetric circuits for compensating the slice level.

2. RF EQUALIZER ARCHITECTURE

The block diagram of the RF equalizer is shown in Figure 1. The signal, picked up by the laser diode assembly, is amplified by a current controlled amplifier (CCA) that sets the signal level to 1Vpp. The equalizer, which uses a gm-C topology, boosts the high-frequency gain using two zeros. This high-frequency boosting is often called as pulse slimming because the pulse looks slimmer after boosting. As is true in the most data communications system, the linear phase characteristic of the filter is of prime interest to maintain the pulse timing integrity. The transfer function of two symmetric zeros, H(S) = (1+As)(1-Bs), is used so that a linear phase characteristic can be obtained. The high-frequency out-of-band noise is attenuated with a -42dB/oct slope.

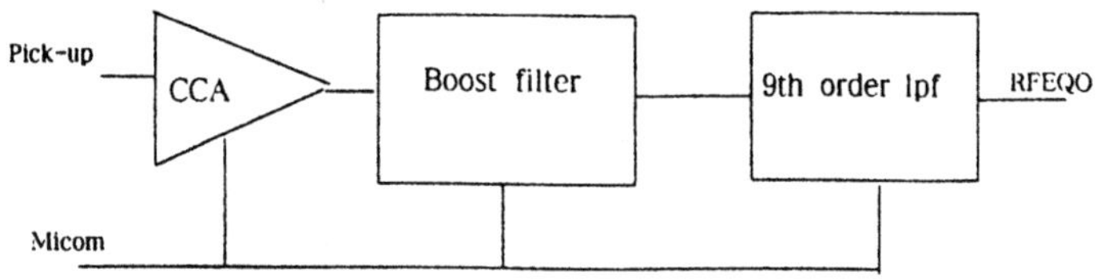

Figure 1. Block diagram of the RF equalizer.

The 9th order Bessel lowpass filter is made of a cascade of 5 biquad stages [1, 2]. All biquad stages are scaled so that all node voltages can have the same magnitude at the cut-off frequency of the equalizer. Figure 2 illustrates the first biquad stage with 2 zeros. It can be implemented fully differentially, but the single-ended version is shown. The tuneable gm is drawn as a resistor in the figure. The cut-off frequency of the equalizer is controlled by gm1 while the boost amount is set independently by gm2 of the frequency control. The tuneable range is set to cover the process variation of about ±10%.

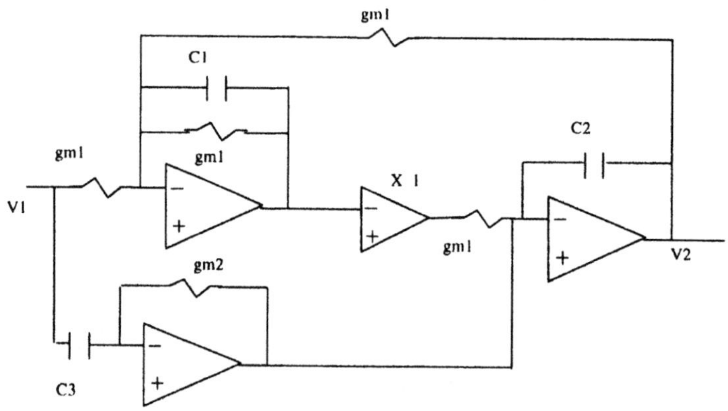

Figure. 2. First biquad with 2 zeros.

The simulation results of the gain and group delay of the DVD equalizer are shown in Figure 3.

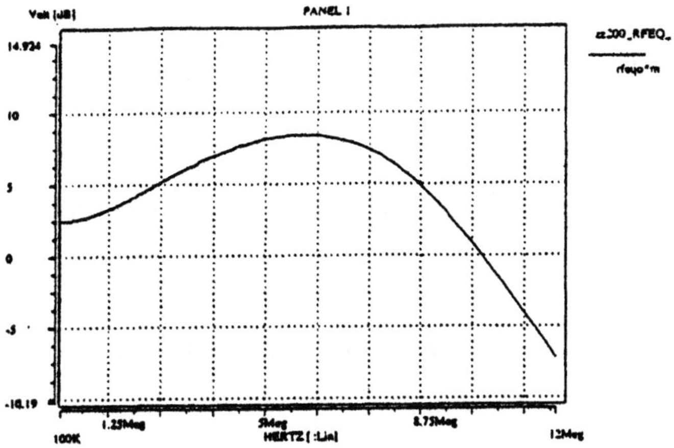

(a) Gain

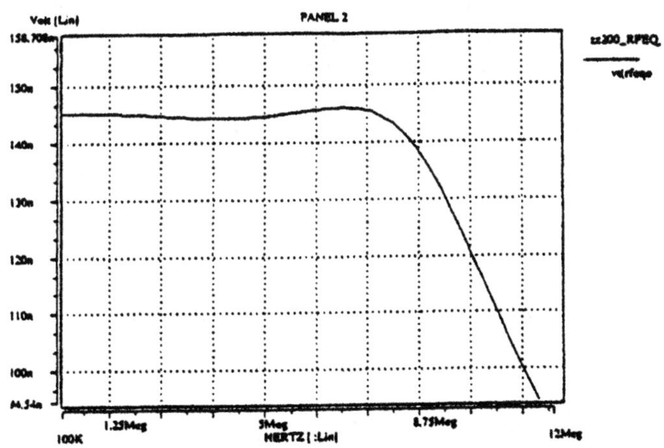

(b) Group Delay

Figure 3. Simulated gain and delay of the DVD equalizer.

The boost and group delay deviation over 1 to 6MHz for DVD are designed to be 6dB at 6MHz and within 2ns, respectively.

3. THE DECISION BLOCK

The decision block diagram is shown in Figure 4. The decision block converts the equalized analog signal to digital data. It affects the performance of the digital signal process as following system.

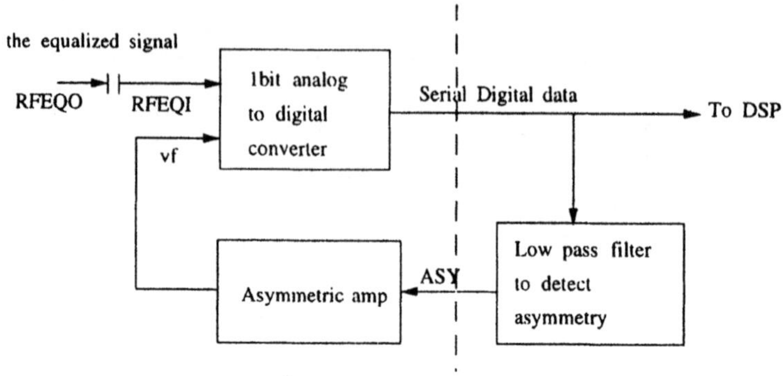

Figure 4. Block diagram of the decision

3.1. One bit analog to digital converter

The block diagram of one bit analog to digital converter is shown in Figure 5. The input is compared with the feedback voltage, vf, that is compensated on the asymmetric error of digital data by the low pass filter and asymmetric amp. The comparator shown in Figure 6 has a hysteresis of about 70 mV to make it insensitive to noise [3]. Also, the main ground and the nmos bulk ground, DGND and AGND, are separated, respectively. It is laid out very carefully to minimize digital noise in analog block.

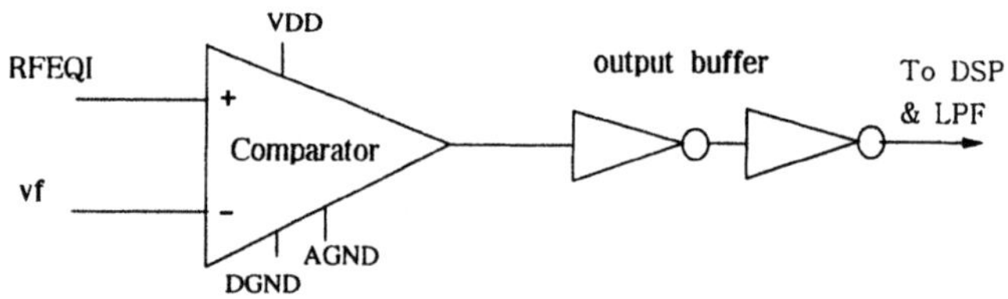

Figure 5. One bit analog to digital converter

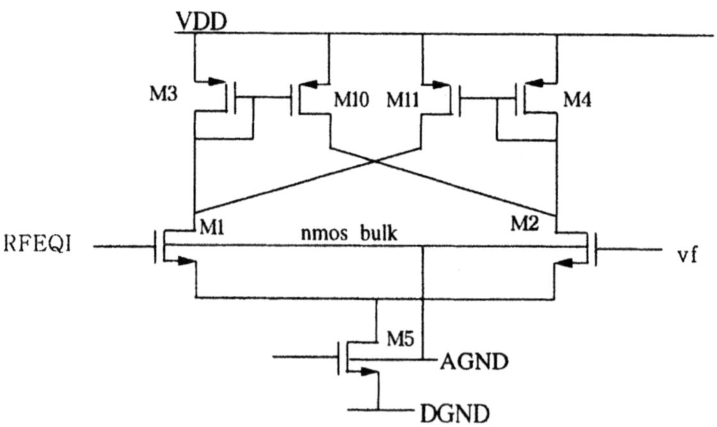

Figure 6. Comparator with a hysteresis.

3.2. Auto asymmetry control amplifier

It is shown in Figure 7 that sets the adequate signal level for the comparator. The reference voltage VC is adjustable to appropriate level to adapt to versatile disc mode.

$$Vout = VC + 80k(VC\text{-}asy)/20k \tag{1}$$

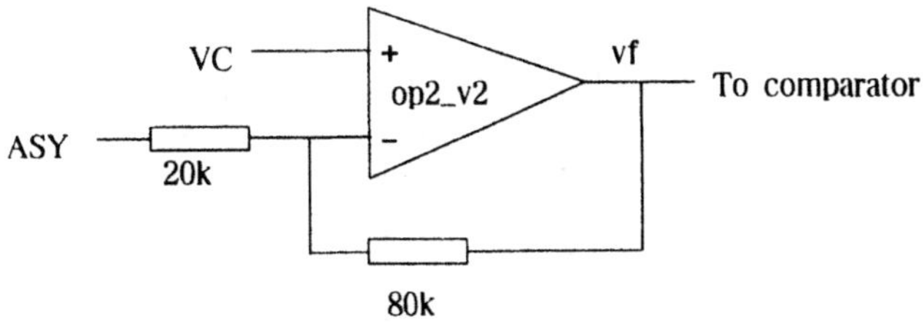

Figure 7. Auto asymmetry control amplifier Circuits

4. EXPERIMENTAL RESULTS

Fabricated using a 0.8um CMOS technology that uses the n-well p-type substrate, the RF amp chip occupies 3000um x 1000um. Measured gain and phase responses of the equalizer shown in Figure 8 are very close to the simulated ones shown in Figure 3.

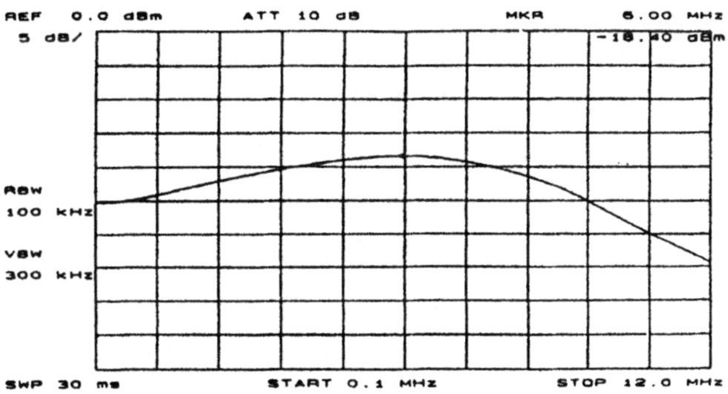

(a) Gain

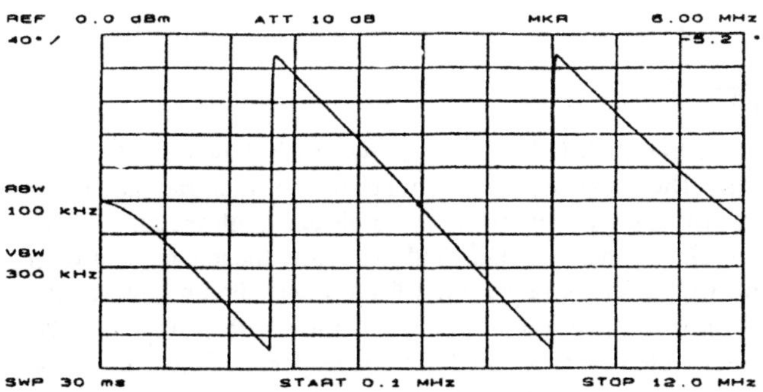

(b) Phase

Figure 8. Measured gain and phase of the equalizer.

When the gain at 6MHz is boosted up to 6dB, the group delay deviation is within 2 to 4ns and the phase linearity is measured to be less than 1%. The jitter measured within 4.5ns is very good for optical disc player. Figure 9 shows the frequency tuning characteristics of the RF equalizer.

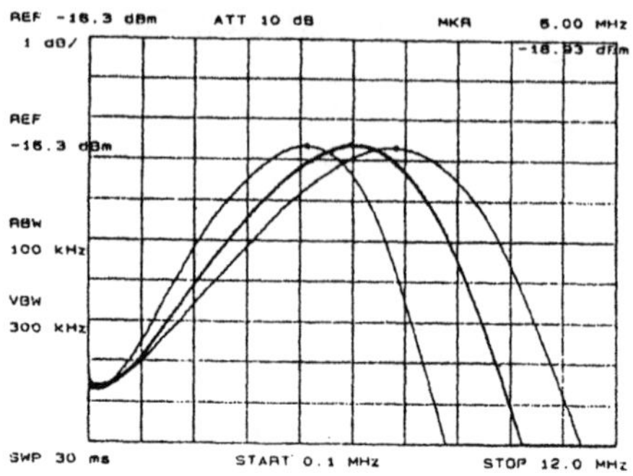

Figure 9. Measured frequency tuning capability.

As designed, the measured tuning range of the cutoff frequency is about ±10%. The cutoff frequency can be set independently of the gain boost. The gain and the frequency of the RF equalizer are set with a Digital-to-Analog converter (DAC) which is controlled by an external microcomputer.

Measured response of the decision block shown in Figure 10 is very symmetrical and any digital noise could not be found in input signal.

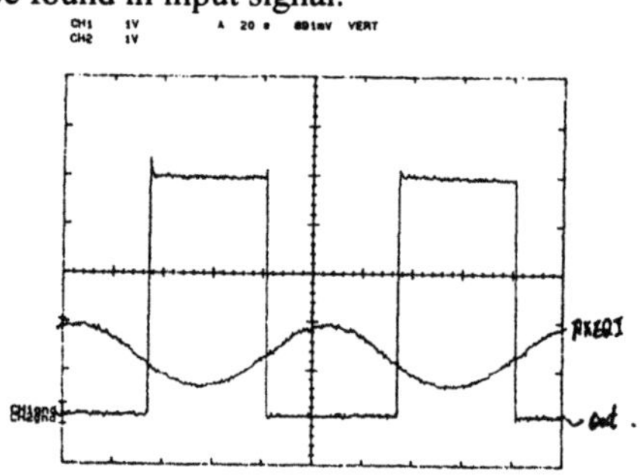

Figure 10. Measured response of the decision block

5. CONCLUSIONS

This paper demonstrates CMOS versions of two basic functional blocks for optical disc playback systems. The equalizer effectively boosts the filter gain at 6MHz up to 6dB and maintains the phase linearity within 1% up to 6MHz. Both the amounts of gain boosting and the filter cutoff frequency are made independently tunable by controlling the gm stage of the equalizer with a tuning range of ±10% to cover the process variation. A 9th order filter with two zeros attenuates high-frequency noise with a slope of -42dB/oct. Finally, the signal jitter characteristic is within 4.5n. The response of the decision block is not only very symmetrical but also does not to affect analog block.

REFERENCES

[1] A.B. Williams and Fred J. Taylor, *Electronic Filter Design Handbook*, 2nd ed., McGraw-Hill, 1988.

[2] R. Schaumann, M.S. Ghausi and K.R. Laker, *Design of Analog Filters:Passive, Active RC, and Switched-Capacitor*, Englewood Cliffs, NJ: Prentice-Hall, 1989.

[3] Phillip E. Allen and Douglas R. Holberg, *CMOS Analog Circuit Design*, Holt, Rinehart and Winston, Inc, 1987.

4

CMOS CURRENT CONVEYOR DESIGN AND MACROMODEL

Stanisław Kuta, Witold Machowski and Robert Wydmański

University of Mining and Metallurgy
Department of Electronics
30-059 Kraków, al. Mickiewicza 30
POLAND

ABSTRACT

The paper presents the design of second generation current conveyor (CCII) with low voltage regulated cascode current mirrors in CMOS technology. An accurate nonlinear macromodel of CCII has also been developed and results of simulation at transistor level model and macromodel are presented and compared.

1. INTRODUCTION

The second-generation current conveyor (CCII) introduced by Sedra and Smith in late '60s has gained wide acceptance as a versatile building block both in theoretical considerations as well as in practice of analog signal processing. CCII may be suitable for different kind of amplifiers, impedance converters and inverters as well as non-linear applications [1,2,3].

Over the past decade many circuit implementations of CCII have been proposed. For some years an emphasis has been laid on solutions appropriate for integration with low voltage digital CMOS circuits. The design presented in this article summarises our former approaches and efforts of making high performance low voltage CCII [4,5].

Conceptually CCII is an active three port (X,Y,Z) network with terminal voltages and currents (to be more specific: their total instantaneous values) obeying the following matrix equation (plus/minus sign denotes noninverting/inverting CCII, respectively):

$$\begin{bmatrix} Iy \\ Vx \\ Iz \end{bmatrix} = \begin{bmatrix} 0 & 1 & 0 \\ 1 & 0 & 0 \\ 0 & \pm 1 & 0 \end{bmatrix} \cdot \begin{bmatrix} Vy \\ Ix \\ Vz \end{bmatrix} \tag{1}$$

Since practically it is impossible to meet all the requirements provided by Equation 1 (real circuit could not have infinite input/output impedance, the bandwidth has to be limited, etc.) this model is almost useless for more advanced analysis. Moreover, Fabre et al. [3] warn: „*some circuits previously published in the literature will only be able to function correctly at low frequency, because the parasitic impedances modify considerably the nature of transient response at high frequencies*". High performance CCII should exhibit simultaneously low input impedance Z_X at port X, very high input impedance Z_Y at port Y and output impedance Z_Z at port Z, possibly widest and flat current- and voltage transfer functions ($\alpha(\omega)=\dfrac{I_Z(\omega)}{I_X(\omega)}$ and $\beta(\omega)=\dfrac{V_X(\omega)}{V_Y(\omega)}$, both being reasonably close to unity at DC) as well as negligible offsets.

The fulfilment of the afore-mentioned requirements in CMOS technology is very often restricted by the voltage input buffer, which is usually unity-gain configured voltage amplifier (a feedback loop from X output to Y input is used in such a case). Introducing the feedback (with the loop gain LG) improves the linearity of static characteristics, guarantees the unity of β as well as causes the decrease of both Z_X impedance and voltage offset by the factor of (1+LG). Although higher LG improves some parameters, increasing of LG causes another problems: big values of gain are frequently achievable only after introduction of the additional second gain stage, which reduces the bandwidth. On the other hand, high gain requires usually big transistor's aspect ratio, which subsequently degrades the noise performance and necessary transistors splitting reduces the bandwidth again. Last but not least, high value of LG is risky due to potential instability of the system. In that way the final design is the result of many reasonable trade-offs.

In the paper we present the design of CMOS CCII with typical architecture but dedicated for low voltage (±1.8V) operation. This feature has been achieved utilizing regulated cascode current mirrors [6].

Current mode signal processing units (like filters, impedance converters, generators rectifiers and other and non-linear applications) utilising CCIIs usually comprise some number of them. It is quite clear, simulation of such a system at the transistor level is very time consuming and may increase convergence problems. In recent years some CCII macromodels have been proposed, which claim simplicity and accuracy [8], nevertheless they exhibit some limitations when an input voltage buffer utilises the op-amp in unity gain feedback configuration. In the paper we present the improved version of the macromodel which very accurately models frequency dependence of Z_X, Z_Y and Z_Z impedance, differential and common mode gain of the input voltage buffer, the output current limitations and voltage and current transfer functions (α,β) in wide frequency range.

2. SCHEMATIC AND LAYOUT

The full circuit diagram of the designed conveyor is presented in Figure 1. All MOS symbols rounded by ovals in the schematic depict the regulated cascode circuits [6] - it is convenient to treat them as a kind of composite transistors (actually each of them is composed of six MOSFETs) possessing low saturation voltage and very high output impedance. Unfortunately due to the positive feedback applied in the composite transistors, very careful sizing and biasing is required in order to achieve an aperiodic step current response.

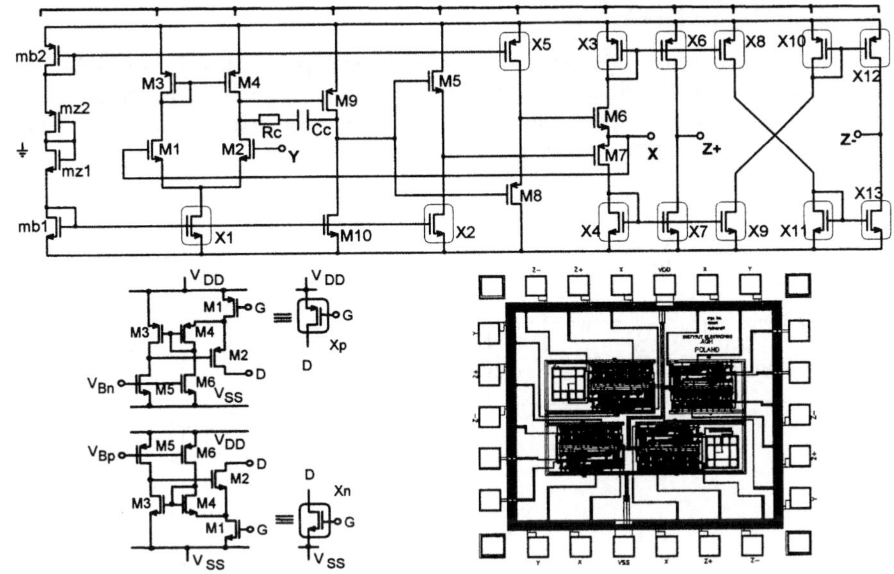

Figure 1. Circuit diagram of proposed CCII (top); the composite transistors (bottom left) and final layout containing four CCIIs with output pads (bottom right)

The voltage follower part of the circuit comprise two gain stages and non- inverting AB push-pull output buffer (M1-M8). The differential pair with an active load forms the input stage. Second gain stage consists of common source amplifier with active load. 100% negative feedback is applied from X to Y. Open loop gain of the amplifier and Miller compensation network elements (R_C and C_C) have been optimized in order to reach the trade-offs mentioned in the introduction, i.e. moderate Z_X, high Z_Z, low output voltage offset and negligible high frequency peaks on α and β characteristics (the overshoot is essentially zero with respect to the voltage transfer function and less than 10% with respect to the current one). Regulated cascode current mirrors (X3-X13) form the noninverting (Z+) conveyor output and after the cross coupling the inverting one (Z-).

The layout has been designed using CADENCE with 1.2 μm AMS CMOS design kit. Main benchmarks obtained from simulation of the extracted layout (the netlist containing all the transistors and parasitics) are collected in Table 1. For convenience of comparison all the graphical characteristics will be presented together with results of simulation at the macromodel level in the next section.

Table 1. Summary of CCII performance parameters

Parameter	Value	Parameter	value	Parameter	value
α_0 (Z-)	1.0005	R_x	0.450 Ω	VF output swing	-2.29-2.02V
α_0 (Z+)	1.0002	L_x	4.2 μH	VF offset	2.27 μV
β_0	0.99945	R_y	10 GΩ	current offset (Z+)	16.5 nA
$\omega_\alpha/2\pi$ (Z+)	75 MHz	C_y	0.94 pF	current offset (Z-)	-4.3 nA
$\omega_\alpha/2\pi$ (Z-)	71 MHz	R_z (Z+)	180 MΩ	C_z (Z+)	1.37 pF
$\omega_\beta/2\pi$	57 MHz	R_z (Z-)	184 MΩ	C_z (Z-)	1.08 pF
Power diss.	7.1 mW				

3. MACROMODEL

The macromodel we have developed is generic although particularly dedicated for CCIIs exploiting a voltage amplifier with differential pair in unity gain connection as an input voltage buffer. We paid special attention to accurate modelling of current and voltage transmitance characteristics over wide frequency range.

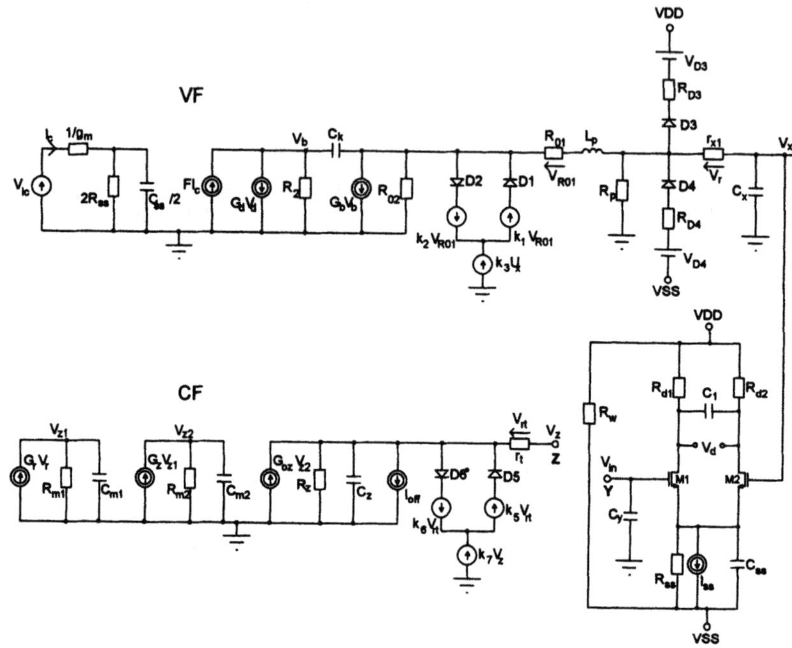

Figure 2. Macromodel of CMOS CCII

Our original invention is an accurate modeling of CMRR - decreasing with frequency, which is a typical feature of amplifier with differential input pair.

The schematic of proposed macromodel is depicted in Figure 2. It consists of two parts: modeling voltage follower (VF) section and current follower (CF) section, respectively. VF section of macromodel is based on Boyle's op-amp macromodel [7], though it is a little improved in order to properly model the CMRR behaviour. Unity gain of the VF is asserted by internal feedback loop from X output to the gate of M2, forming with M1 a differential input pair. G_d and G_b conductance as well as R_2 and R_{02} resistance model differential gain. Two controlled current sources, connected via C_k, simulate intermediate gain stage with dominant pole compensation. By changing the threshold voltage of single transistor from differential pair we model the voltage offset. R_W resistance is utilised for calculating the total dissipated power.

Current source FI_c is used for modelling common mode gain, which increases with frequency. Since in modelled VF follower very strong feedback is applied, impedance seen from the X terminal has not only the poles but zeros as well. Thus there is a necessity of accurate modelling of that impedance, especially in high frequency range, and elements L_p, R_p and C_x are used for that purpose.

CF section is modelled by the second part of the macromodel. Controlled current source G_rV_r repeats current flowing through sampling resistance r_{x1}. $R_{m1}C_{m1}$ time constant determines a dominant pole of the frequency characteristics and $R_{m2}C_{m2}$ introduces the second pole necessary for accurate modelling of current transmitance α.

Appropriate choice of macromodel parameters allows accurate simulation of conveyor characteristics as well as properties of complex linear and non-linear systems containing CCIIs.

Selected characteristics obtained by simulation of the extracted layout as well as proposed macromodel are compared in Figure 3. Almost identical results obtained for both simulations prove excellent accuracy of the macromodel. In the similar way, i.e. using either extracted netlist or macromodel, we simulated some advanced applications of CCIIs current mode rectifiers, impedance converters and generators. As an example, the simulation results of Wien-bridge CCII based oscillator proposed in [10] is presented in Figure 4. It is apparent from Figure 4 both extracted layout and macromodel simulation give similar results of excitation process. The transient responses are slightly different, but steady-state oscillations have the same frequency, amplitude and THD (1.3%).

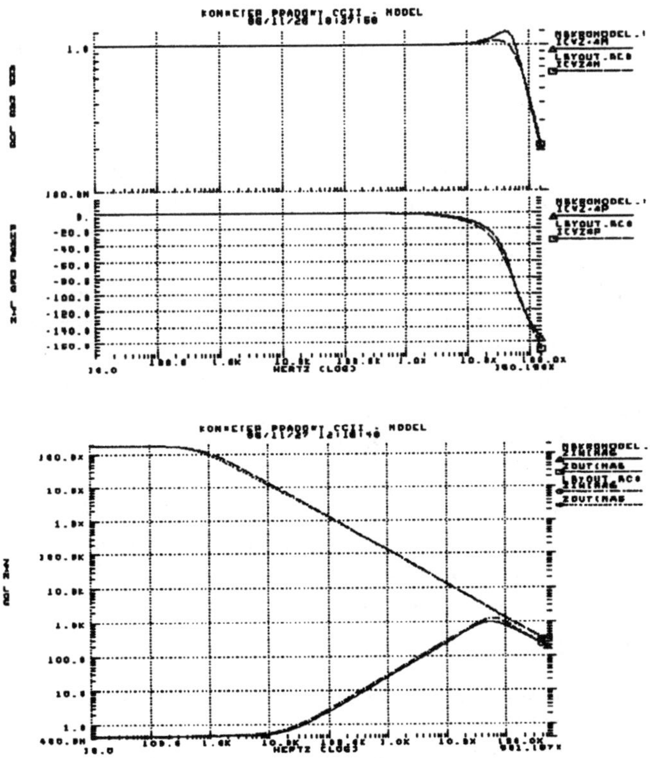

Figure 3. Simulated AC characteristics of α (top), Z_x and Z_z (down) of extracted layout and macromodel.

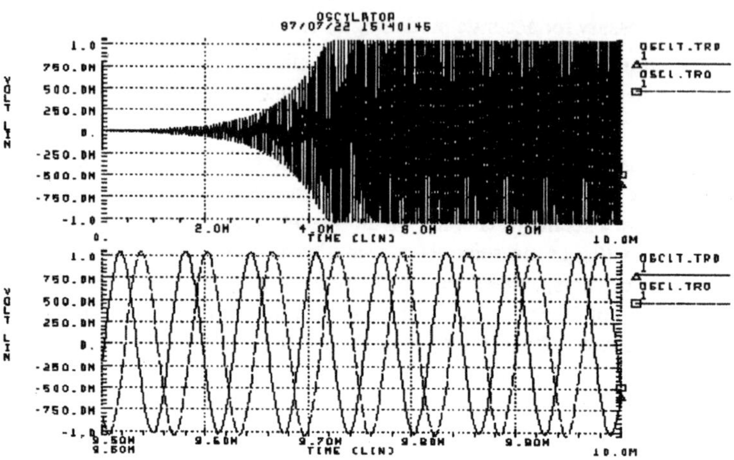

Figure 4. Simulation of excitation process in CCII based Wien oscillator at extracted layout and macromodel level.

4. CONCLUSIONS

Presented circuit implementation of CCII is based on typical architecture with input voltage buffer composed of two stage voltage amplifier configured in unity gain connection and dominant pole frequency compensation. The circuit performances are comparable to that for CMOS CCII solutions known in the literature, though low voltage (±1.8V) operation is achieved by using regulated cascode current mirrors. Their application requires an optimal dimensioning of MOSFETs comprising composite transistors in order to achieve an aperiodic step response of considered current mirrors. As a result, the high frequency overshoot of transfer function $\alpha(\omega)$ is minimized.

Since HSpice simulation results were promising, the chip containing 4 CCII± circuits has been submitted for fabrication via EUROPRACTICE MPW service.

The presented macromodel is comprehensive and accurate both for DC and AC parameters. Its main enhancement is proper modeling the HF decrease of CMRR in the input voltage buffer. The benefit of using the macromodel instead of full equivalent circuit is significant reduction of the simulation time - by the factor of 7 for small-signal analysis and up to 53 for transient analysis of oscillator.

REFERENCES

[1] B. Wilson, *Recent developments in current conveyors and current mode circuits,* IEE Proc. G, Vol. 137, No. 1, pp.63-77, 1990

[2] S.I. Liu, D.S Wu, H.W Tsao, J. Wu and J.H Tsay, *Nonlinear applications with current conveyors,* IEE Proc. G, Vol. 140, No. 1, pp. 1-6, 1993;

[3] A. Fabre, O. Saaid and H. Barthelemy, *On the Frequency Limitations of the Circuits Based on Second Generation Current Conveyors,* Analog Integrated Circuits and Signal Processing, vol. 7, No. 2, pp.113-129, 1995

[4] S. Kuta, W. Machowski and J. Jasielski, *Improved CMOS implementation of current conveyors,* Kwartalnik Elektroniki i Telekomunikacji, Vol. 40, pp. 201-214 (in Polish), 1994

[5] S. Kuta, W. Machowski and J. Jasielski, *CMOS Current Conveyor with Regulated Cascode Circuits for Low Supply Operation* Proc. XVIII National Conference Circuit Theory and Electronic Circuits, Polana Zgorzelisko, pp. 97-102 , 1995

[6] A.L. Coban and P.E. Allen, *A 1.75V rail-to-rail CMOS op amp*, Proc. IEEE ISCAS'94, London, vol. 5, pp.479-500, 1994

[7] G.R. Boyle, P.M. Cohn, D.O. Pedersen and J.E. Solomon, *Macromodeling of Integrated Circuit Operational Amplifiers*, IEEE Journal of Solid-State Circuits, vol.. 9, pp.353-363, 1974

[8] N. Tarim, B. Yenen and H. Kuntman, *Simple and Accurate Nonlinear Current Conveyor Macromodel*, Proceedings -8th Mediterranean Electrotechnical Conference ,MELECON'96, 13-16 May 1996, Bari, Italy, vol.1, pp.447-450, 1996

[9] B. Wilson, *Performance Analysis of Current Conveyors*, Electronics Letters, vol.25, pp.1596-1597, 1989

[10] P.A. Martinez, S. Celma and I. Gutierez, *Wien -Type Oscillators using CCII+*, Analog Integrated Circuits and Signal Processing, Vol. 7 No. 2, pp. 139-148, 1995

5

BEHAVIOURAL NOISE MODELLING OF CROSS-COUPLED RING OSCILLATORS

Leszek J. Opalski

Institute of Electronic Fundamentals
Faculty of Electronics and Information Technology
Warsaw University of Technology
ul. Nowowiejska 15/19, 00-665 Warszawa
POLAND

ABSTRACT

The paper presents a method of behavioral noise modeling for double cross-coupled CMOS ring oscillators, that is suitable for noise analysis and design. The model does not directly use time waveforms, but time moments when corresponding waveforms of each ring cross the threshold level. Dependence of period jitter on size of cross-coupling transistors is also modeled for possible noise optimization of the oscillator.

1. INTRODUCTION

Recent trends to build fully integrated CMOS radio [1,3,5] created interest in low-power consumption low-phase noise components, e.g. GHz frequency range CMOS integrable oscillators. A novel type of ring oscillator (later referred to as the Cross-Coupled Double Ring Oscillator or CCDRO) was introduced in [2,6]. Measurements of experimental implementations in CMOS technology demonstrated ability of the CCDRO circuit to deliver output signal with lower phase noise than it was possible for Single Loop Ring Oscillator (SLRO). So far, no satisfactory theoretical justification of the noise superiority of the CCDRO w.r.t SLRO has been published. This paper fills the gap, as it provides theoretical and numerical justification of the noise properties of CCDRO circuits.

2. CROSS-COUPLED DOUBLE RING OSCILLATORS

Conceptually CCDRO is composed of M cross-coupled inverters (CCI), connected in a loop (see Figure 1a). In practical implementations, reported so far, $M = 3$ and the cross-coupling has been used only for the first two pairs of inverters (the last inverter pair has extra transistors that implement delay/frequency control). One possible implementation of CCI is shown in Figure 1b.

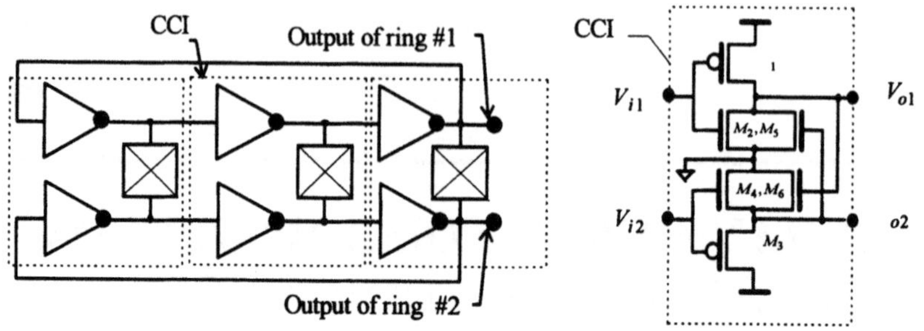

(a) (b)

Figure 1. a) Conceptual schematic of the Cross-Coupled Double Ring Oscillator (CCDRO) with M = 3 Cross-Coupled Inverters (CCI). b) Schematic of the CCI subcircuits.

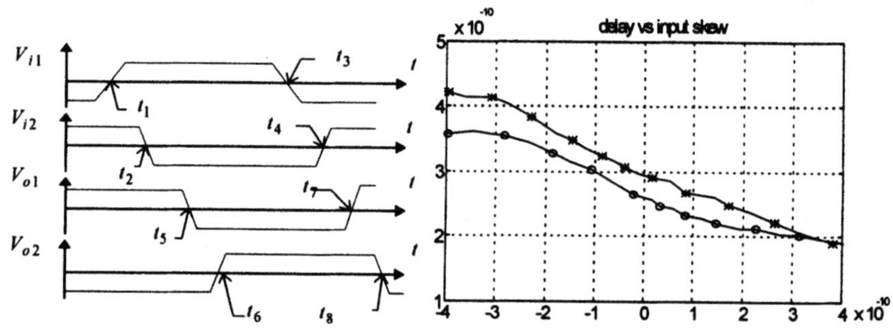

(a) (b)

Figure 2. a) Input/output waveforms and notation of time events in CCIs. b) Example dependence of CCI delay ($V_{i1} \to V_{o1}$) on the two input waveform skews: $\tau_{1,1}(s_{in,1})$ - marked with circles; $\tau_{2,1}(s_{in,2})$ - marked with asterisks.

Let us denote delays of CCI inverters (M_1, M_2 and M_3, M_4 in Figure 1b) as $\tau_{1,1} = t_5 - t_1$, $\tau_{1,2} = t_6 - t_2$, $\tau_{2,1} = t_7 - t_3$, $\tau_{2,2} = t_8 - t_4$ (see Figure 2a). Ideally, inverters of each CCI should be switched by input waveforms in the same time points (but in opposite directions). Switching skews of driving waveforms are denoted as $s_{in,1} = t_1 - t_2$ (for rising V_{i1}) and $s_{in,2} = t_3 - t_4$ (for falling V_{i1}). The coupling transistors (M_5, M_6 in Figure 1b) modify delay times of inverters such, that switching skew of the output waveforms (V_{o1}, V_{o2}) is reduced w.r.t. skew at their inputs. Figure 2b shows an example dependence of inverter delays on input waveform skews for a design in a 1.2 μm CMOS technology. This property of CCI allows for the two coupled rings in CCDRO to synchronize themselves. It will be shown, that the property also makes reduction of noise in the output voltage possible.

3. MODELLING OF JITTER

In the following analysis it is assumed that:

- **[A1]** : all inverters of the CCDRO are identical;
- **[A2]** : inverters propagate square waveforms;
- **[A3]** : noise is small, so propagation delays can be determined from the following linear model:

$$\begin{aligned}\tau_{1,1}(s_{in,1}) &= T_a + \delta\tau_{1,1}(s_{in,1}) &&= T_a - a\cdot s_{in,1} + r_{1,1}\\ \tau_{1,2}(s_{in,1}) &= T_b + \delta\tau_{1,2}(s_{in,1}) &&= T_b + b\cdot s_{in,1} + r_{1,2}\\ \tau_{2,1}(s_{in,2}) &= T_b + \delta\tau_{2,1}(s_{in,2}) &&= T_b - b\cdot s_{in,2} + r_{2,1}\\ \tau_{2,2}(s_{in,2}) &= T_a + \delta\tau_{2,2}(s_{in,2}) &&= T_a + a\cdot s_{in,2} + r_{2,2}\end{aligned} \qquad (1)$$

where $r_{i,j}$ are random variables representing delay jitter. T_a and T_b are nominal (zero input skew) inverter delays for falling and raising V_{i1}, respectively. $a \geq 0$ and $b \geq 0$ are "jitter gain factors" that represent slopes of delay on skew dependencies (such as shown in Figure 2b). To get a closed-form description of jitter the noise sources $r_{i,j}$ of inverters are assumed Gaussian and independent with zero mean and standard deviation σ_r.

From the assumption **A2** it follows (as in [4]) that the output waveforms can be characterized by timepoints t_i at which inverters switch their output (low to high or vice versa). Since distances between these timepoints are determined by direction of waveform changes and values of CCI delays, as determined from equations (1), in what follows we will also refer to the switching moments via their indices (in square brackets).

Let us analyze a CCDRO with M CCIs. Assume that at a timepoint t_{n-m} inputs of the first CCI cause the upper inverter to start switching. Outputs V_{o1}, V_{o2} of the first CCI change after delays of $\tau_{1,1}[n-m]$ and $\tau_{1,2}[n-m]$, respectively. This triggers the second CCI, and so on. The initial switching at inputs to the first CCI propagates to the second, up to the M-th CCI - forming the first part of the periodic output. The change propagates back to the first CCI, but with opposite switching direction than before. After next M consecutive delays the changes appear again at the CCDRO outputs - thus commencing the period of oscillations. The period of the output waveforms can be therefore expressed as:

$$\begin{aligned}T_{1,m}[n] &= \tau_{1,1}[n-m+1]+\tau_{2,1}[n-m+2]+\ldots+\tau_{1,1}[n-1]+\tau_{2,1}[n] =\\ &= \underbrace{MT_0}_{T} + \underbrace{\sum_{k=1}^{M}(\delta\tau_{1,1}[n-m+2k-1]+\delta\tau_{2,1}[n-m+2k])}_{\delta T_{1,m}[n]} \qquad (2)\\ T_{2,m}[n] &= T + \underbrace{\sum_{k=1}^{M}(\delta\tau_{1,2}[n-m+2k-1]+\delta\tau_{2,2}[n-m+2k])}_{\delta T_{2,m}[n]}\end{aligned}$$

where T_0 is an average inverter delay, $m = 2M$, and arguments in brackets determine indices of timepoints. It is numerically advantageous to consider only the jitter of periods $\delta T_{i,m}[n]$, $i = 1,2$. After some symbolic algebra manipulation we get:

$$\delta T_{1,m}[n] = \sum_{k=0}^{m-1} c_{1,k}(r_1[n-m+k]-r_2[n-m+k]) + \sum_{k=1}^{m} r_1[n-m+k]$$
$$\delta T_{2,m}[n] = \sum_{k=0}^{m-1} c_{2,k}(r_2[n-m+k]-r_1[n-m+k]) + \sum_{k=1}^{m} r_2[n-m+k] \qquad (3)$$

$$c_{i,k} = -u_i\frac{1-z^{m-k}}{1-z^2} \qquad \text{for k = 0,2,4,...,m, } i = 1,2$$
$$c_{1,k} = -b+u_1 z\frac{1-z^{m-k-1}}{1-z^2} \qquad \text{for k = 1,3,5,...,m - 1} \qquad (4)$$
$$c_{2,k} = -a+u_2 z\frac{1-z^{m-k-1}}{1-z^2} \qquad \text{for k = 1,3,5,...,m - 1}$$

where $z = a+b$, $u_1 = a-bz$, $u_2 = b-az$. Obviously, statistics of jitter at the end of waveform period (index $[n]$ in the formulae above) depend also on switching jitter of the beginning of the whole period (index $[n-m]$). For the following noise calculation statistical steady state was assumed, which means that statistics of jitter of the beginning and the end of the same period are equal.

Symbolic calculations of jitter statistics are lengthy and very tedious. The results obtained are particularly compact, easy to read and to analyse, when additional symmetry of switching is assumed for all inverters (i.e. $T_b = T_a = T_0$, $b = a$). After symbolic manipulation of the equations (3-4) we arrive at the following set of linear equations w.r.t. $V = E\{(\delta T_1)^2\}$, $C = E\{\delta T_1 \delta T_2\}$.

$$\begin{bmatrix} 2c_{1,0}^2 & 1-2c_{1,0}^2 \\ 1+2c_{1,0}^2 & 2c_{1,0}^2 \end{bmatrix} \cdot \begin{bmatrix} C \\ V \end{bmatrix} = \sigma_r^2 \cdot \begin{bmatrix} 1+\sum_{k=1}^{m-1}(1+2c_{1,k}+2c_{1,k}^2) \\ 2\sum_{k=1}^{m-1}(1+c_{1,k})c_{1,k} \end{bmatrix} \qquad (5)$$

Figure 3 shows dependence of the normalized variance $V_r = V/V_0$ and correlation coefficient of the output waveform periods $\rho = C/V$ on the jitter gain factor a of the linear noise model (1) for a range of values of $m = 2M$ (M being the number of CCIs). Since $V_0 = m\sigma_r^2$ denotes variance of periods generated by the corresponding SLRO with the same inverters as used in CCDRO and the same number of stages, the normalized variance V_r is a measure of noise advantage of the CCDRO over SLRO.

It is clearly seen that for $a \in (0,1/2)$ CCDRO exhibits lower period jitter than the corresponding SLRO, with maximum jitter variance reduction $\approx 30\%$. It is also seen that both curves depend weekly on the total number of stages used (M).

From dependence of the correlation coefficient on the gain factor a it can be also inferred that the noise advantage of the double ring structure is caused by positive correlation of noisy switching that occurs in both rings. The correlation is created by cross-coupling of the two rings.

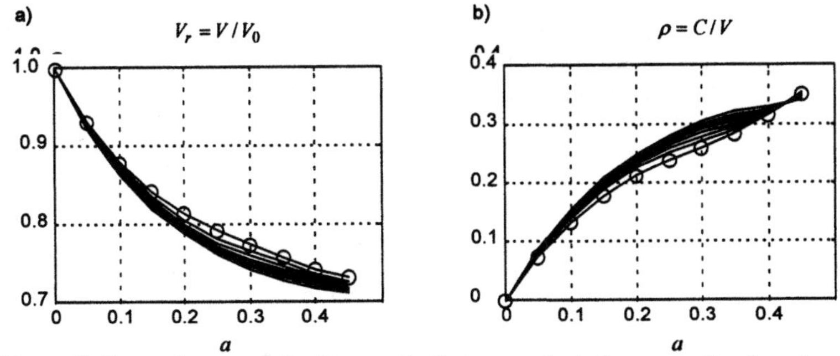

Figure 3. Dependence on the jitter gain factor a of: a) the normalized variance $V_r = V / V$ *and b) the correlation coefficient* $\rho = C / V$ *of CCDRO output waveform periods, for m=6,8,...,20, as determined from the linear jitter propagation model. The curve corresponding to m=6 is marked with circles.*

4. BEHAVIOURAL NOISE MODELLING

Noise oriented design requires modeling of jitter dependence on designable parameters. In what follows minimum size transistors are assumed (for maximum frequency); only gate width w of the coupling transistors (M_5 , M_6 in Figure 1b) is assumed designable.

Behavioral noise modeling of CCIS consists of 3 stages. First, for each gate width w the following generic behavioral delay model is fitted to data from HSPICE simulations for a range of input skews s_{in} :

$$\tilde{\tau}(s_{in}, x) = c(x_1 \tanh(x_2 \frac{s_{in} + cx_3}{c}) + x_4) \tag{6}$$

where x is a vector of model parameters and c a scaling constant (e.g. 1e-10 for a 1.2 μm CMOS technology). Model parameters x depend on the designable parameter w. Actual form of dependence of x vectors on designable parameters is assumed, based on the range of allowable values of parameter w and observed accuracy of approximation. The dependence can be assumed e.g. of the form:

$$\begin{aligned} x_1(z,w) &= \max(z_1 \cdot w + z_2, z_3 \cdot w + z_4) \\ x_2(z,w) &= \min(z_5 \cdot w + z_6, z_7 \cdot w + z_8) \\ x_3(z,w) &= \min(z_9 \cdot w + z_{10}, z_{11} \cdot w + z_{12}) \\ x_4(z,w) &= \max(z_{13} \cdot w + z_{14}, z_{15} \cdot w + z_{16}) \end{aligned} \quad \text{for } \tau_{1,1}, \tau_{2,2} \tag{7}$$

$$\begin{aligned} x_1(z,w) &= z_1 \cdot \tanh(z_2 \cdot w + z_3) + z_4 \\ x_2(z,w) &= \max(z_5 \cdot w + z_6, z_7 \cdot w + z_8) \\ x_3(z,w) &= z_9 \cdot \tanh(z_{10} \cdot w + z_{11}) + z_{12} \\ x_4(z,w) &= z_{13} . \tanh(z_{14} \cdot w + z_{15}) + z_{16} \end{aligned} \quad \text{for } \tau_{1,2}, \tau_{2,1} \tag{8}$$

At the third modeling step models (6-8) are combined and resulting models fitted again. For a 1.2 μm CMOS process design and range of w up to 6 μm the fit was excellent: the mean square relative error was below 0.03% for input skews as shown in Figure 2b.

The behavioral noise model of CCIs was used for estimation of the period jitter according to the linear theory of sec. 1.3, as well as to Monte Carlo based calculations. In Figure 4 a dependence of variance of the period jitter on the designable parameter w is plotted - as calculated by Monte Carlo sampling using full nonlinear models (6-8). A minimum can be found by inspection or by a separate optimization process. The optimum value from Figure 4 is very close to the best value that has been found by a designer for that particular CMOS process.

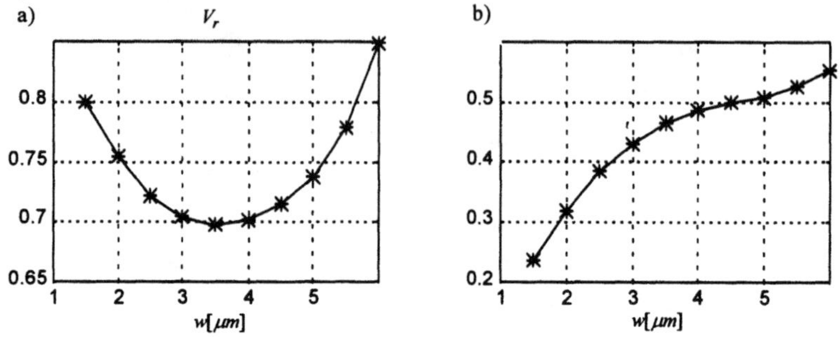

Figure 4. Monte Carlo estimates of: a) dependence of scaled variance of period jitter V_r and b) dependence of correlation coefficient ρ of output waveforms periods on designable parameter w.

5. CONCLUSIONS

The results presented in this paper ascertain that the noise advantage of CCDRO over SLRO, that has been observed in measurements of experimental CMOS circuits, is due to the structural differences of the two types of generators. The novel structure of CCDRO creates positive correlation between jitter of its two sub-rings, which in effect reduces jitter of both output waveforms. A behavioural noise modelling methodology is also introduced, and illustrated - showing its value for design process.

The author is grateful to Prof. Kwasniewski from Carleton University, Ottawa, Canada for inspiration, generous support and cooperation during initial stage of this research. Help with HSpice simulation of CCDRO from Leszek Zyra is also acknowledged.

REFERENCES

[1] P.R. Gray and R.G. Meyer, *Future Directions in Si ICs for RF Personal Communication*, Digest of Technical Papers, CICC, May 1995, Santa Clara, CA.

[2] T.A. Kwasniewski, M.Abou-Seido, A. Bouchet, F. Gaussorgues and J. Zimmerman, *Inductorless oscillator design for personal communication devices - A 1.2 µm CMOS process case study*, Digest of Technical Papers, CICC, May 1995, Santa Clara, CA, 327-330.

[3] T.A. Kwasniewski, M.Abou-Seido and A.J. Bergsma, *Towards an all CMOS Digital Radio*, Proc. of the 7th Int. Conf. on Wireless Communications, Wireless 95, 10-12 July 1995, Calgary, Alberta, CA, 115-120.

[4] J. McNeil, *Jitter in Ring Oscillators*, Proc. ISCAS'94, London, 201-204.

[5] M. Thamsirianunt and T.A. Kwasniewski, *A 1.2 µm CMOS Implementation of a Low-Power 900 MHz Mobile Radio Frequency Synthesizer*, Digest of Technical Papers, CICC'94, May 1994, Santa Clara, CA, 383-386.

[6] M. Thamsirianunt and T.A. Kwasniewski, *CMOS VCOs for PLL Frequency Synthesis in GHz Digital Mobile Radio Communications*, Digest of Technical Papers, CICC'95, May 1995, Santa Clara, CA, 331-334.

[7] T.C. Weigandt, B. Kim and P.R. Gray, *Analysis of Timing Jitter in CMOS Ring Oscillators*, Proc. ISCAS'94, London, 1994, 27-30.

6

A 27MHZ FULLY-BALANCED OTA-C FILTER IN 2μM CMOS TECHNOLOGY

Bogdan Pankiewicz, Jacek Jakusz and Stanisław Szczepański

Faculty of Electronics Telecommunications and Informatics
Technical University of Gdańsk
ul. G. Narutowicza 11/12, 80-952 Gdańsk
POLAND

ABSTRACT

This paper presents a second-order CMOS continuous-time filter tuneable from 7 to 27MHz cutoff frequency. A linear, fully-balanced, voltage-tunable CMOS operational transconductance amplifier (OTA) with large DC gain and wide bandwidth is also described. The approach uses a two-differential-pair transconductor with a cross-coupled input stage together with a negative resistance load for compensating the parasitic output resistance of the OTA. Since no additional internal nodes are generated, DC gain enhancement is obtained without bandwidth limitation. This amplifier is used to design a second-order lowpass OTA-C filter in the high-frequency range, fabricated in a standard 2μm n-well CMOS process through MOSIS. The measured filter response is very close to the SPICE simulated response.

1. INTRODUCTION

The integration in CMOS technologies of filters in the megahertz range is very attractive for video, IF, and other applications. Among integrated continuous-time filters, the transconductance-capacitor (OTA-C) filter is one important implementation method which, in recent years, has been successfully applied in CMOS technology to design filters with frequency range up to 100MHz [2]-[6]. In this paper, the design of a lowpass OTA-C filter in 2μm CMOS technology with cutoff frequency tunable in a range 7-27MHz is described.

In the Section 2, the design of a fully-balanced OTA which employed a negative resistance load for obtaining high dc-gain and wide bandwidth is presented. The design of a second-order lowpass OTA-C filter is presented in Section 3. Finally, in the last section, the simulated and measured results are summarized.

2. OPERATIONAL TRANSCONDUCTANCE AMPLIFIER

In OTA-C filters, there are only two types of components: OTAs and capacitors. Therefore, the design of an OTA with the required performance is a very important step in OTA-C filter implementation. A CMOS OTA based on two cross-coupled differential nMOS pairs M1, M2 and M3, M4 is shown in Figure 1. The identical MOS devices

M1-M4 which form voltage to current converter are operating in saturation, and both pairs are biased by a DC current sink (Mzb6) in combination with a floating voltage source V_B, connected between nodes ***3*** and ***4***. In the range of operation the dc current of Mzb6 is assumed constant and tuning is achieved by adjusting V_B. A detailed description can be found in [1, 7]. Using the standard square-law model for MOS devices the transconductance of the stage is given by:

$$gm = \frac{\partial I_{OUT}}{\partial V_{id}} = 2k_n V_B \tag{1}$$

where V_{id} is the differential input voltage, I_{OUT} is the differential output current and $k_n = 0.5\mu_n C_{ox} W / L$ is the transconductance parameter of devices M1-M4. All undefined parameters have their usual meaning. Thus, the transconductor stage exhibits perfectly linear transconductance *gm*. In practice, due to second order effects, additional devices M1a-M4a are added to improve the linearization of DC transfer function and for high-frequency response compensation as well [1].

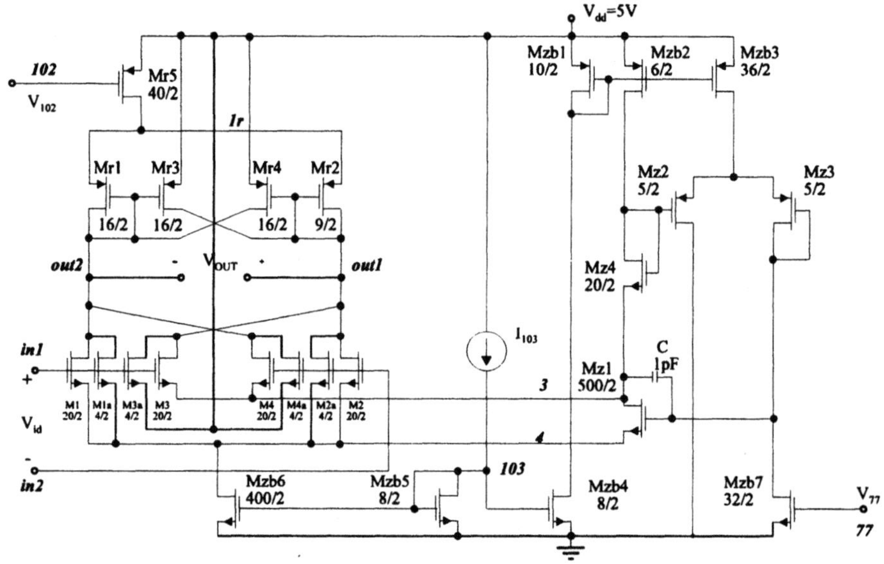

Figure 1. Complete CMOS OTA circuit diagram.

Voltage controllable negative resistance load (NRL) is formed by devices Mr1-Mr5. It is used for compensating of parasitic output resistance of MOS devices. The proposed NRL structure does not have any extra signal nodes and thus has very good frequency response. Negative resistance appears only between output nodes ***out1*** and ***out2*** while the common-mode (CM) load resistance is relatively low and due to this no common-mode feedback (CMFB) circuit is necessary [1], [7]. For stability of stand alone OTA a very precise control of negative resistance value by varying the DC voltage V_{102} is required. Fortunately in most application of these building blocks, such as analog filters, stability is easier to obtain because of interaction with other elements in complete structure [3].

The non-ideal floating DCc voltage source V_B is composed of devices Mz1-Mz3, Mzb1-Mzb3 and Mzb7. It is a simple voltage shifter with output transistor Mz1 which is made relatively large for obtaining low output resistance of the source. Voltage V_{77} is used for controlling of the output voltage V_B between nodes *3* and *4*. An additional transistor Mz4 is used for extending the tuneability and for temperature compensation of the proposed DC voltage source. The complete OTA operates with a single 5-V voltage supply and its power consumption is 3.5mW.

The OTA was laid out with the MAGIC tool. Figure 2 shows the layout of the OTA from Figure 1. An outline shape is almost square (208μm x 195μm). An ohmic contacts were placed around OTA for isolating purposes. The guard rings were placed around n-well to prevent CMOS latch up. Capacitance *C* was made as POLY1/POLY2. The V-I converter and NRL subcircuits are placed nearby and have compact layout. Input and output pins are placed on the opposite sides.

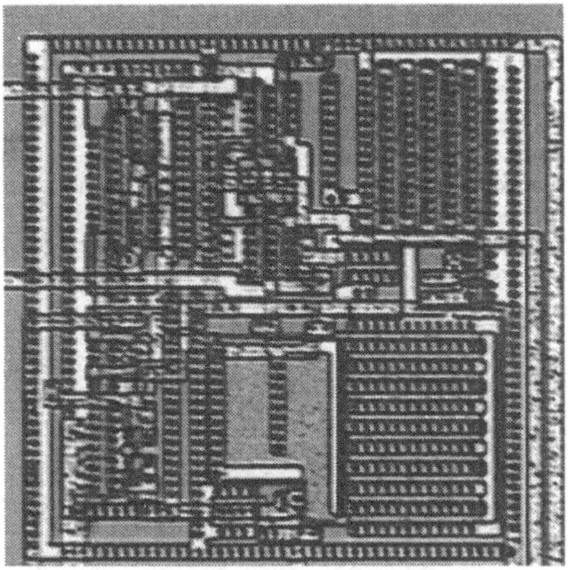

Figure 2. Microphotograph of the OTA in Figure 1.

3. FILTER IMPLEMENTATION

For testing of the OTA, second-order lowpass OTA-C filter with quality factor Q=2 has been designed and fabricated. Electrical scheme of this filter is presented in Figure 3. An additional OTA which works as a buffer is added at the end of the filter (not shown in Figure 3), so that the output signal is measured in current mode. This is made to avoid influence of IC pins capacitance on frequency filter response. Transfer function of a biquadratic filter is given by:

$$\frac{V_{OUT}(s)}{V_{IN}(s)} = \frac{\omega_o^2}{s^2 + s\frac{\omega_o}{Q} + \omega_o^2} \qquad (2)$$

where: $\omega_o = gm/\sqrt{C_1 C_2}$ and $Q = \sqrt{C_2/C_1}$.

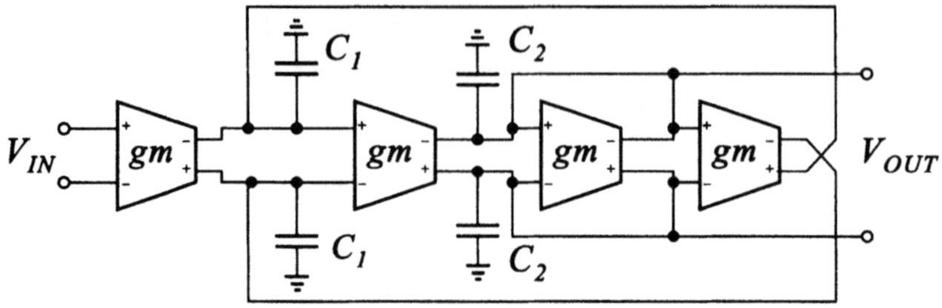

Figure 3. Circuit diagram of the implemented second-order lowpass filter.

It can be seen from equation (2) that the poles of this filter are proportional to *gm* and inversely proportional to C_1 and C_2. Therefore, in order to achieve a higher cutoff frequency for the filter, values of C_1 and C_2 should be small and *gm* should be large. Since each OTA of the filter in Figure 3 has parasitic input and output capacitances parallel to C_1 and C_2, the values of C_1 and C_2 have a lower limit: they should be larger than parasitic capacitances (by factor ≥ 5) to reduce effect of unpredicted process tolerances and maintain predictability [6]. In the proposed filter architecture, all parasitic capacitances are at nodes where circuits capacitors are located. Thus, to minimize this effect, the filter capacitances are predistorted in the design by subtracting all parasitic from the nominal capacitances used in filter. The final values used in design are C_1=1,286pF C_2=5,001pF, parasitic capacitance is less than 20%.

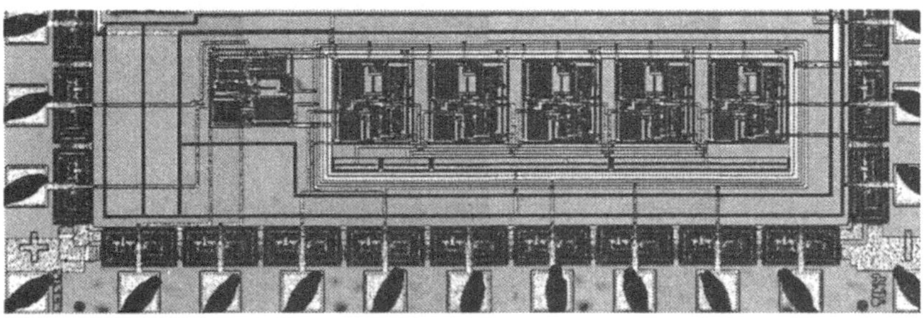

Figure 4. Microphotograph of the filter from Figure 3.

Layout of the filter is presented in Figure 4. Capacitors were built as POLY1 over POLY2. To eliminate the parasitic capacitance from POLY1 to substrate, the POLY1 layer was connected to the ground. Therefore all capacitors C_1 and C_2 in the layout are grounded. To reduce the wire resistance, the shortest routing was used. All OTAs are isolated from another by surrounding ohmic contacts.

4. SIMULATION AND EXPERIMENTAL RESULTS

This section contains simulation results and experimental results of the OTA and the filter. They were fabricated through MOSIS in 2μm n-well technology. The chip was measured using MARCONI PF 2370 spectrum analyzer with 300Ω load and digital DC multimeters. Figure 5 shows close agreement between simulated and measured DC transfer characteristics of the OTA. Measured transconductance range is 40-125μS (simulated 35μS -130μS).

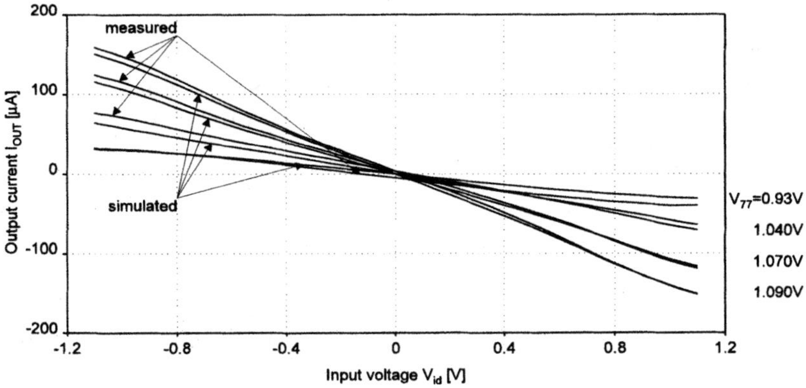

Figure 5. Simulated and measured DC transfer characteristic of the OTA in Figure 1.

Measured and simulated filter frequency responses are presented in Figure 6. The 3dB cutoff frequency is tuned in the frequency range 6.9MHz-27.1MHz (simulated 6.83-25.4MHz).

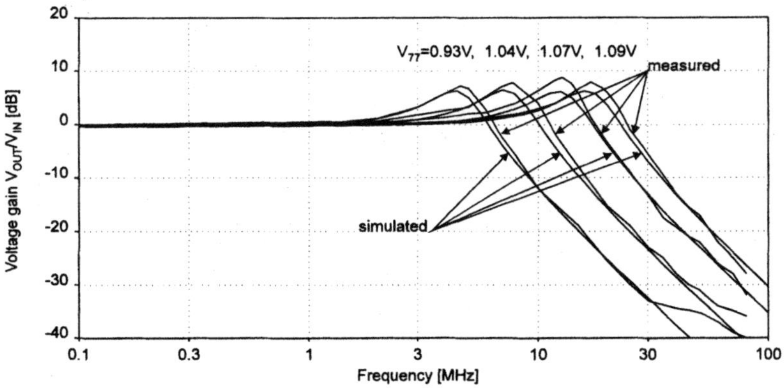

Figure 6. Simulated and measured frequency response of the filter in Figure 3.

Measured total harmonic distortion (THD) for 5.5MHz input sine wave are presented in Figure 7. THD is lower than -40dB for signal as large as $320mV_{pp}$.

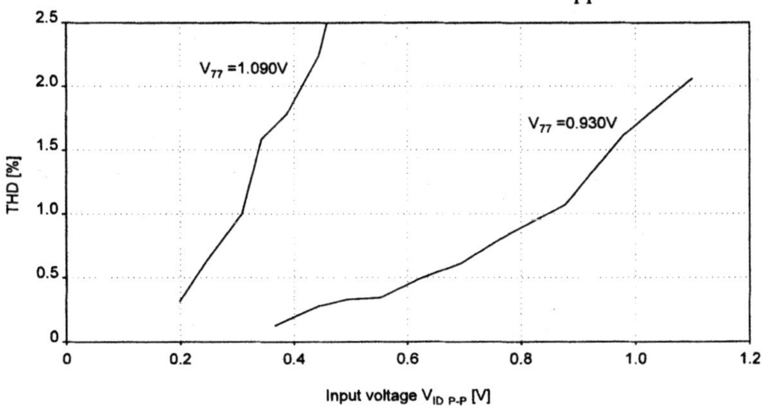

Figure 7. Measured THD of the filter in Figure 3

5. CONCLUSIONS

A 7 to 27MHz lowpass second-order OTA-C filter chip has been presented. The filter with quality factor equal to 2 has been fabricated in 2μm n-well CMOS technology through MOSIS. Measurements are close to simulations using LEVEL 2 SPICE parameters. The chip uses single supply voltage of only 5V and power consumption is 3.5mW per one OTA. THD is lower than -40dB for signal as large as $320mV_{pp}$. Area of the OTA is 208μm x 195μm and of the filter is 1192μm x 360μm.

REFERENCES

[1] S. Szczepański, J. Jakusz and R. Schaumann, *A Linear Fully-Balanced CMOS OTA For VHF Filtering Applications,* accepted for publication in IEEE Transactions on Circuits and Systems-II, (scheduled to appear in CAS-II, Vol. 44, No. 3, March 1997).

[2] H. Khorrambadi and P.R. Gray, *High-Frequency Continuous-Time Filters,* IEEE J. Solid-State Circuits, Vol. 19, No. 6, pp. 939-948, Dec. 1984.

[3] B. Nauta, *Analog CMOS filters for very high frequencies,* Kluwer Academic Publishers, 1993

[4] J.M. Khoury, *Design of a 15-MHz CMOS Continuous-Time Filter with On-Chip Tuning,* IEEE J. Solid-State Circuits, Vol. 26, No. 12, pp. 1988-1997, Dec. 1991.

[5] J. Silva-Martinez, W.S.J. Steyaert and W. Sansen, *A 10.7-MHz 68-dB SNR CMOS Continuous-Time Filter with On-Chip Automatic Tuning,* IEEE J. Solid-State Circuits, Vol. 27, No. 12, pp. 1843-1853, Dec. 1992.

[6] P. Wu, R. Schaumann and R. Daasch, *A 20 MHz Fully-Balanced Transconductance-C Filter in 2μm CMOS Technology,* in Proc. IEEE International Symposium on Circuits and Systems, pp.1188-1191, 1993.

[7] S. Szczepański, J. Jakusz and B. Pankiewicz, *VHF Linear Fully-Balanced CMOS OTA with 5V Power Supply* in Proc. XVII National Conference of Circuit Theory & Electronic Circuits, Poland, vol.2, pp. 675-680, Oct. 1994.

7

PROGRAMMING ANALOG NON-VOLATILE MEMORIES

Eric Tournier and Jean-Louis Noullet

Laboratoire d'Analyse et d'Architecture des Systèmes du CNRS
7, av. du Colonel Roche - 31077 Toulouse, Cedex 4
FRANCE
Fax: (33) 61 33 62 08 - email: tournier@laas.fr

ABSTRACT

Analog memories are devices which can store any analog value taken from a given range, at the opposite of digital memories, for which only two values are represented. The values are stored in voltage form, by modulating the quantity of charges at a given point and thus the voltage influence of it at a measured point. The analog programming of such memories requires a precise control of charge, to have a precise control of potential, and is by this aspect far more difficult to implement than digital programming. Another difficulty of analog memories implementations concerns the retention quality. The analog stored value is directly concerned by parasitic modification of charges, whereas the digital stored value can be refreshed periodically to eradicate the parasitic influences.

1. INTRODUCTION

The fantastic technological effort in industry for making electronic memories is driven by the digital needs. If you are doing a full analog circuit and need to store a value, the only choice you have by now is to use a digital memory with A/D and D/A converters to write it in a digital way, and read it. To integrate the whole on a single chip, there is a need of compatible technology, and a need of many extra area for the A/D and D/A converters. You can decide to use two chips, the second being specific for digital memorizing, but you loose in integration density. Both cases result in extra investments. The storage in analog form beside the circuit would then be very attractive. Furthermore, the existence of analog memories could lead to a novel interest in analog implementations. For example in neural networks, analog implementations would be far more compact than their digital counterparts [1].

2. THE ANALOG MEMORIES

It seems safe to start from digital memories implementations to deduce possible analog memories implementations. Among digital memories, only the (E)EPROM can offer analog capabilities. The Figure 1 shows the principle of EEPROM. Charges are stored on a piece of polysilicon surrounded by insulated material. Modification of charges results from Fowler-Nordheim tunnelling [2] through insulating oxide under the influence of a high electric field. This complete insulation of charges defining the potential to store is able to ensure analog retention, whereas non-(E)EPROM implementations with no complete insulation are not sufficient for analog retention. EEPROM are generally implemented as floating gate MOS (a control gate and a floating gate) [3]. But single poly EEPROM can also be done [4].

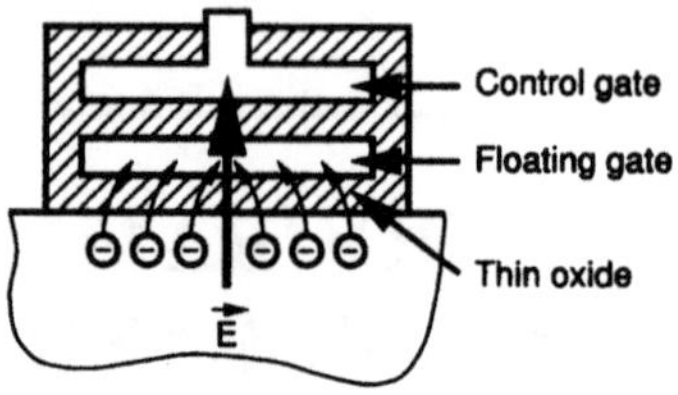

Figure 1. EEPROM principle

Unlike in digital, an accurate control of charges in the analog writing must be achieved. The main problem is that the writing need a high electric field through the structure, provided by a capacitive coupling, for Fowler-Nordheim tunnelling to take place. This capacitive coupling alters the floating gate potential to be measured. Thus, measuring at the same time than writing leads to a different value of that found when not writing. Some solutions have been proposed [5]. Programming can be done by dissociating write and measurement: a succession of write without measurement (blind write) and measurement without write (no capacitive coupling alteration). The purpose is to adjust the value step by step, beginning with big changes and finishing with small (depending on the accuracy needed). The resulting complication for the circuit is not negligible, because of the non-real time sensing: a measurement system to evaluate the difference with the target after each pulse, a system which supervises the succession of steps (e.g. a sequencer or a small micro controller), leading to a digital control for the analog value. The advantage towards the digital memories is then much less obvious. Moreover, specific device of technology can be required [6].

We can keep an analog point of view for charge control. A new scheme, conciliating capacitive coupling and simultaneous sensing, leads to a simplify closed loop configuration.

3. AN ANALOG MEMORY MODELLING

Two capacitances C_1 and C_2 share a common floating polysilicon electrode (Figure 2). Suppose that $V_e=0$, Q_{tot} is the charge on the floating gate, and C_{tot} is the capacitance seen from it. The stored potential is then Q_{tot}/C_{tot} relative to ground. The total potential on the gate is given by adding the capacitive coupling ratio $V_{cc}=C_1/(C_1+C_2)\ V_e$ during write.

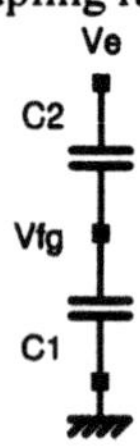

Figure 2. Capacitive divider

You cannot distinguish the two contributions. You can decide to neglect the coupling ratio by maximizing C_2/C_1: C_2 reach very big value and thus would take too much area on chip (C_1 cannot be lowered unlimitedly). At the opposite, you can decide to take the capacitive coupling into account. ANEEP estimates the capacitive coupling, and adds it to the target value. A comparator between this new value and V_{fg} is used to detect the end of write. A difference between capacitances values, $C_1 >> C_2$, is necessary for two reasons:

- if $C_1 = C_2$, both capacitances see the same coupling: there are two tunnel currents which cancel themselves. If $C_1 \neq C_2$, The smallest capacitance see the highest coupling and then injects the highest tunnel current. If C_2 is the main injector (i.e. the smaller), writing increases V_{fg}.
- if $C_1 >> C_2$, V_{cc} potential ratio due to coupling stays close to the mass and consequently, the total potential V_{fg} can be sensed with a MOS gate (because of insulation) without oxide breakdown during write.

4. READING

An operational amplifier (OA) follower gives the value V_{fg} (Figure 3). The PMOS follower needed to sense V_{fg} voltage on the (+) pin is duplicated on the (-) pin. By paying attention to the design, both can be matched together. To read small values, the low supply of the OA must be pull down to a negative value, so that the transistors of the output stage can work properly. This constraint appears only for reading small values.

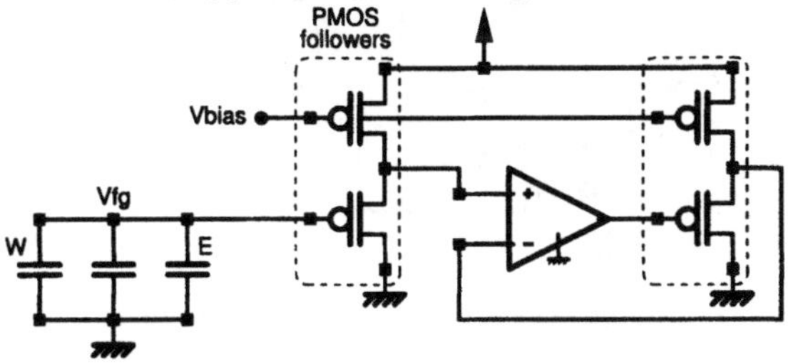

Figure 3. Reading part of the circuit

5. WRITING

Two steps are needed for write control (Figure 6): firstly the capacitive coupling estimation, and secondly the real time comparison with target value (allowed by the first point). The capacitive coupling $V_{cc} = C_1/(C_1+C_2)\ V_{pwr}$ can be estimated by a resistance divider $V_{ccr} = R_1/(R_1+R_2)\ V_{pwr}$ (Figure 4) and added to the target value (V_g), through an OA (the divider and the adder are merged as shown in Figure 6). V_{cc} as V_{ccr} are based on relative values (ratios), allowing good accuracy. The comparison is done between $V_g + V_{ccr}$ and $V_{fg} + V_{cc}$ (Figure 5), the low supply of the reading OA is the mass, so that the minimum value readable is about 0.5 V). Reading lower values is simply done by using a symmetric supply, whereas the memories remains referenced to the mass.

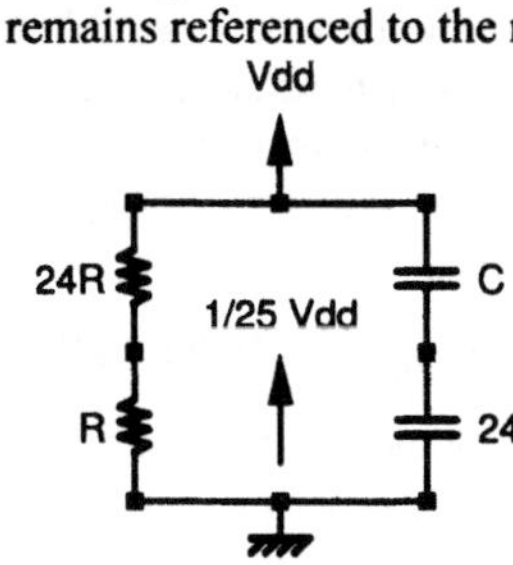

Figure 4. Resistance divider equivalent to capacitive divider

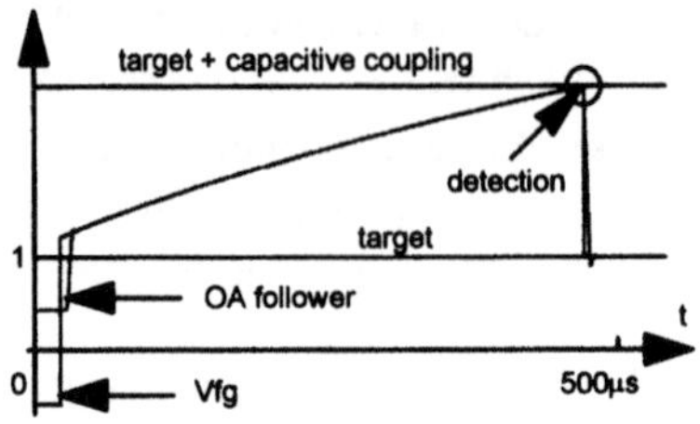

Figure 5. Write and detection of the good value

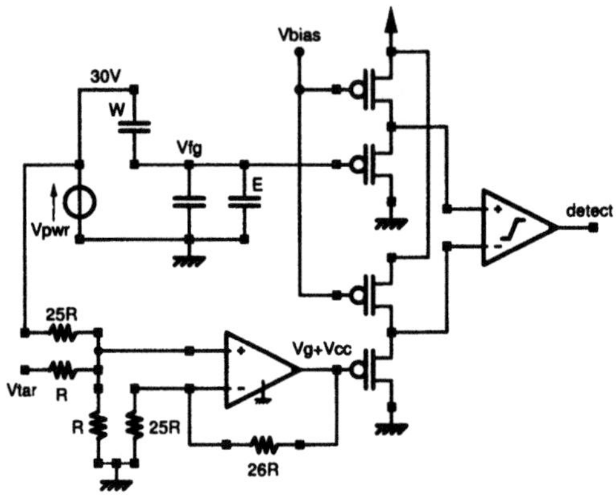

Figure 6. Writing part of the circuit

6. ERASING

No control of erasing is done. When erasing goes on, a negative coupling is applied so that V_{fg} is pulled down. An approximate end is commanded by a comparator, the threshold V_t of which is: V_{gs}-$|V_{cc}|$-ε (offset from PMOS follower, negative coupling, security margin on the final value). When detecting the end of erasing (V_{fg} decreases $\Rightarrow$ $V_{fg}=V_t$), the negative coupling is suppressed (V_{fg} increases), with the result that the comparator does not detect the end any more ($V_{fg}>V_t$), and the erasing restarts again. To cancel ending oscillations of such a circuit in a simple way, a second comparator with a higher threshold ($=V_{gs}$) is used to detect the need of erasing. The final value for an erased memory is therefore -ε.

7. SUPERVISING RUN MODES

A simple supervising logic has been integrated. Two SR latches are used to keep in mind the end of write and the need of erase (these are two digital memories). The main reason for the first is to avoid oscillations at the end of write with the lost of coupling (like in erase mode), and for the second is that once erase is begun, the value is not intentional any more and should be completely erased.

8. LAYOUT OF THE ANALOG MEMORY

A prototype was fabricated on the Smart-Power Mietec HBIMOS 2µm technology. The use of Smart-Power technology was justified because we wanted to switch the voltage needed for tunnel effect (≈ 30 *V*) directly on the chip, whatever it was (no external circuit is required). Furthermore, Smart-Power technologies know a growing interest. Yet, a BICMOS analog technology should be suitable to integrate the needed "high voltage" drivers as soon as the oxide breakdown voltage (high limit for tunneling voltage) can be switched with PN junctions. For technologies with limited voltage, external drivers can be used. A precedent work has shown how the number of drivers can be drastically reduced, limiting the number of connections with the chip [7].

Figure 7. Layout of ANEEP

The tunneling through poly1-poly2 oxide was shown to be easier than through poly-diffusion oxide, even with a higher thickness [8]. In other words, we may conclude that poly-diffusion oxide should offer better retention than poly1-poly2 oxide, at the expense of a more difficult write. The chip has 12 essentials inputs/outputs plus 19 others for testing purpose. There are 8 memories. The whole layout is 2.5 × 3 mm^2 (Figure 7). Each extra memory adds about 0.09 mm^2 on the total area.

9. ELECTRICAL TESTS RESULTS

For this first circuit, our main goal was to prove that the analog use of EEPROM with a simple control of charge was feasible. As a consequence, a little prejudice on design has been found at first measurements: the capacitive divider is not 1/25 as expected but 1/17.75, whereas the resistance divider shows the correct factor value 1/25. This implies a new way of reading the results, but with no influence on the conclusions. This mismatch leads to a resulting error of (1/17.75-1/25) Vpwr from the target value.

Assume a target of 2.483 V, and Vpwr=29 V for a slow write (a few seconds). The calculated error is (1/17.75-1/25) 29 V=0.474 V. The resulting target value is then 2.483 V-0.474 V=2.009 V. Here are the results for the eight memories of three different ANEEPs. [tmin ; tmax] represents the interval of time necessary to achieve the write for the fastest and the slowest memory on the same chip.

	1	2	3	4	5	6	7	8	$[t_{min};t_{max}]$(s)
ANEEP 1	1.995	1.995	1.996	2.000	1.995	2.003	1.998	1.997	[5 ; 30]
ANEEP 2	2.002	2.009	2.010	2.003	2.008	2.005	2.011	2.003	[20 ; 120]
ANEEP 3	2.005	2.009	2.010	2.012	2.002	2.011	2.003	2.005	[0.2 ; 0.5]

The different memories respond differently to V_{pwr}, not only between different chip, but between memories of the same chip too: from 0.2 second to 120 seconds. Nevertheless, the final values well match each others. Inside a chip, the maximum dispersion is about 10 mV, and is observed between the slowest and fastest memories: we believe this to be an influence of the time response from the comparator to the switch, during which the tunneling continues: Figure 9 shows the error versus V_{pwr} voltage, that is versus the speed of write: The faster the write, the higher the error. It is the compromise speed/accuracy. Differences between ANEEPs comes from their own comparator and OA offsets. The error with the target is about 15 *mV*. On a [0 ; 5 *V*] range, a precision of 20 *mV* is "8-bits equivalent". The range can be extended to 8 *V* for V_{dd}=*12 V*.

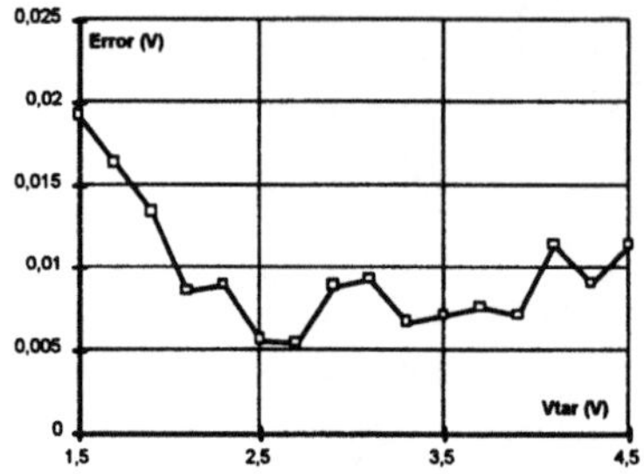

Figure 8. Accuracy versus Vtar

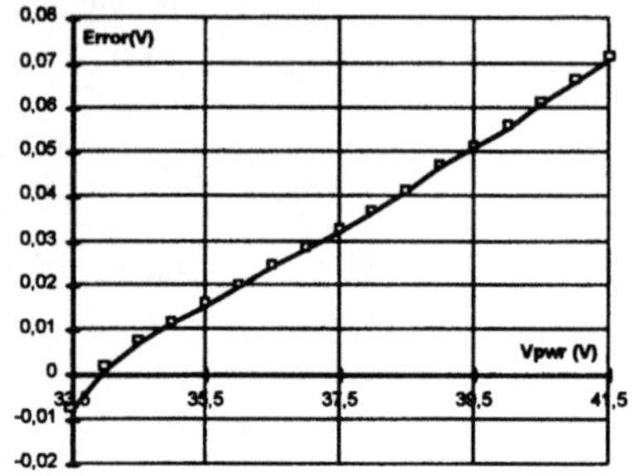

Figure 9. Accuracy versus Vpwr

The accuracy on the global range for the same chip is shown in Figure 8. The evolution is due to non-linearity of diffusion capacitances used. The use of double polysilicon capacitances would cancel it.

10. IMPROVEMENTS

As we use OA and comparator with MOS inputs for writing and reading, with relatively high input offset which diminish the accuracy, we can make use of floating gate MOS to cancel this offset [9]. The chip can be use to program the needed thresholds. This will be a self-correcting circuit: its first work will be to "memorize" its own parameters.

REFERENCES

[1] D.D. Caviglia, M. Valle and G.M. Bisio, *A VLSI module for analog adaptative neural architectures*, Elsevier Science Publishers B. V. (North-Holland), 1992.

[2] E.H. Snow, *Fowler-Nordheim tunneling in* SiO_2 *films*, *Solid State Communications*, vol. 5, pp. 813-815, 1967.

[3] W.S. Johnson, G.L. Kuhn, A.L. Renninger and G. Perlegos, *16-k EE-PROM relies on tunneling for byte-erasable program storage*, *Electronics*, pp. 113-117, February 1980.

[4] K. Ohsaki, N. Asamoto and S. Takagaki, *A single poly EEPROM cell structure for use in standard CMOS processes*, *IEEE Journal of Solid-State Circuits*, vol. 29, pp. 311-316, March 1994.

[5] M. Hooler, S. Tam and R. Benson, *An electrically trainable artificial neural network (ETANN) with 10240 floating gate synapses*, *Proc. Int. Joint Conf. Neural Networks*, vol. II, pp. 191-196, 1989.

[6] O. Fujita and Y. Amemiya, *A floating gate analog memory device for neural networks*, *IEEE Transactions on Electron Devices*, vol. 40, pp. 2029-2035, November 1993.

[7] J.L. Noullet, E. Tournier and A. Ferreira, *Analog non-volatile memory cells for use in ASICs*, in *First Ibero American Microelectronics Conference X'SBMicro I'IberMicro, Canela (Brazil)*, August 1995.

[8] D.A. Durfee and F.S. Shoucair, *Comparison of floating gate neural network memory cells in standard vlsi CMOS technology*, *IEEE Transactions on Neural Networks*, vol. 3, pp. 347-353, May 1992.

[9] E. Säckinger and W. Guggenbühl, *An analog trimming circuit based on a floating-gate device*, *IEEE Journal of Solid-State Circuits*, vol. 23, pp. 1437-1440, December 1988.

8

FOUR-QUADRANT CMOS AMPLIFIER FOR LOW-VOLTAGE CURRENT-MODE ANALOG SIGNAL PROCESSING

Ryszard Wojtyna, Piotr Grad and Jarosław Majewski

University of Technology and Agriculture
Institute of Telecommunication
ul. Kaliskiego 7, 85-763 Bydgoszcz
POLAND

ABSTRACT

The paper presents a low-voltage current-mode CMOS amplifier, whose gain can be electronically controlled over a range covering both negative and positive values. Hence, it can be called four-quadrant amplifier, by analogy to four-quadrant multipliers. Like multipliers, the amplifier can be used, among others, to weighting synaptic connections in artificial neural networks. To improve its linearity, a novel cascode mirror has been proposed. Both small and large signal theoretical studies are given. SPICE simulation results presented confirm the theoretical predictions.

1. INTRODUCTION

Rapid progress in CMOS technology means decreasing all dimensions of MOS transistors and a necessity to reduce supply voltages. Current-mode circuits are much better suited for low supply conditions than voltage-mode ones. Designing low-voltage analog circuits is more difficult than designing digital circuits. To fully exploit the low supply voltages, the circuit structure should consist of many current paths connected in parallel between the positive and negative supply rails. In each of these paths no more than one gate-source voltage V_{GS} should occur when summing voltages along the path. This is because $|V_{GS}|$ of a MOS transistor is relatively high and, typically, is much higher than a minimum value of $|V_{DS}|$, necessary for the transistor to operate in saturation. Moreover, it is obvious that the absolute value of the transistor threshold voltage $|V_T|$ should be as low as possible. Low values of $|V_T|$ can be easily obtained in sub-micron CMOS processes. However, in digital circuits $|V_T|$ is deliberately made to be high (close to 1V) in order to maintain a sufficient noise margin. Thus, the parallel structure of an analog circuit and the requirement that no current path between the supply rails should include more than one voltage V_{GS} is of particular significance in case of mixed-mode (analog-digital) low-voltage IC's.

Weighting synaptic connections in analog artificial neuron networks can be realized by means of multipliers or amplifiers. In [1], an electronically controlled current mode multiplier has been proposed, which is characterized by a low power consumption. Disadvantages of the circuit of [1] are complexity and a poor linearity. Current-mode amplifiers suitable for weighting synaptic connections have been proposed in [2]. Unfortunately, negative and positive gains in the amplifiers of [2] can be achieved at two different outputs. Here, a current-mode amplifier is proposed, where, according to a control voltage V_{TUN}, both negative and positive gains can be achieved at a single output.

2. BASIC STRUCTURE OF THE AMPLIFIER AND ITS SMALL-SIGNAL PROPERTIES

The amplifier circuit is shown in Figure 1. It consists of 5 parallel current paths between the supply rails. In none of these paths there is more than a single gate-source voltage V_{GS}. The voltages V_{GS} appear in these paths due to the transistors M5, M2 and M7. The idea of achieving both negative and positive gains at a single output is realized by subtracting two gain coefficients, one of which, $G(V_{TUN})$, is changeable, and the other, β, is constant. From Figure 1 it is seen that the output current I_{OUT} is a sum of drain currents of M9 and M8 minus the drain currents of M10 and M4. This leads to:

$$I_{OUT} = I_s - \beta I_{IN} \tag{1}$$

As a consequence, the amplifier gain is given by:

$$A = \frac{I_{OUT}}{I_{IN}} = \frac{I_s}{I_{IN}} - \beta = G(V_{TUN}) - \beta \tag{2}$$

To determine the variable component $G(V_{TUN})$ of the overall gain A, notice that the transistors M1, M2, M5, M6, M11 form an input transconductor while the transistors M3, M4, M7, M8, M12 an output transconductor. The input transconductor converts the amplifier input current I_{IN} into a voltage $V_{D2}=V_{G2}=V_{G3}$. By means of the output transconductor, this voltage is next converted into the current I_S.

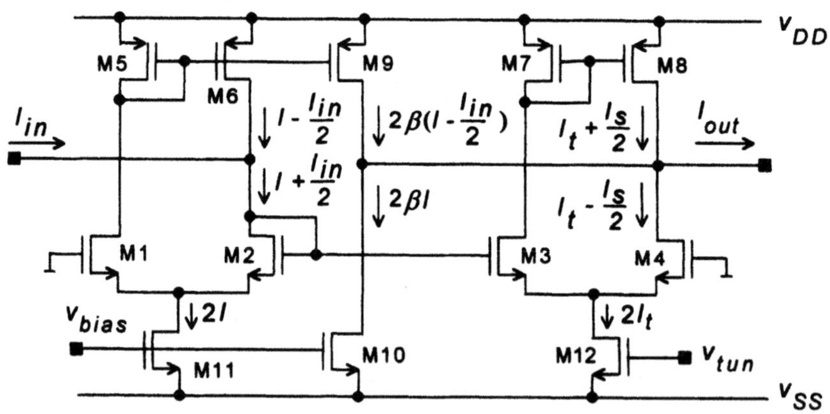

Figure 1. Basic structure of the proposed low-voltage current-mode CMOS amplifier

Denote by g_{mo} transconductance of the output transconductor and by g_{mi} that of the input one. Taking into account the square-low characteristic of a MOS transistor:

$$I_D = K(V_{GS} - V_T)^2, \tag{3}$$

the variable current gain of (2) can be expressed as:

$$G(V_{TUN}) = \frac{I_S}{I_{IN}} = \frac{g_{mo}}{g_{mi}} = \sqrt{\frac{K_o}{K_i}}\sqrt{\frac{I_t(V_{TUN})}{I}}\ , \tag{4}$$

where K_o relates to the differential pair M3, M4 and K_i to the pair M1, M2.

From (4) it is seen that the gain $G(V_{TUN})$ can be varied either by changing the tail current $2I$ of the input transconductor or by changing the tail current $2I_t$ of the output one. In the proposed amplifier one makes use of the second possibility.

3. LARGE SIGNAL PROPERTIES

For the amplifier shown in Figure 1, the following equation can be written:

$$V_{GS2} - V_{GS1} = V_{GS3} - V_{GS4} \tag{5}$$

Drain currents of the transistors M1, M2, M3 and M4 in Figure 1 are described by:

$$I_{D1} = I - \frac{I_{IN}}{2};\quad I_{D2} = I + \frac{I_{IN}}{2};\quad I_{D3} = I_t + \frac{I_S}{2};\quad I_{D4} = I_t - \frac{I_S}{2} \tag{6}$$

Substituting an inverse function of (3) into (5) and taking into account (6) one obtains:

$$\sqrt{\frac{I + \frac{I_{IN}}{2}}{K_i}} - \sqrt{\frac{I - \frac{I_{IN}}{2}}{K_i}} = \sqrt{\frac{I_t + \frac{I_S}{2}}{K_o}} - \sqrt{\frac{I_t - \frac{I_S}{2}}{K_o}}\ , \tag{7}$$

provided that threshold voltages of the transistors M1, M2, M3, M4 are the same.

For K_i and K_o being equal, solving (7) for I_S yields the following nonlinear relation between the I_S component of the output current I_{OUT} and the input current I_{IN}:

$$I_S = 2I_t\sqrt{1 - \left(1 - \frac{I}{I_t}\left[1 - \sqrt{1 - \left(\frac{I_{IN}}{2I}\right)^2}\right]\right)^2} \tag{8}$$

For positive values of I_{IN}, the relation (8), normalized with respect to the tail current I of the differential pair M1-M2, takes the shapes shown in Figure 2. Notice that the function $I_S = f(I_{IN})$, given by (8), is linear only when $I = I_t$.

SPICE simulations of the amplifier shown in Figure 1 were performed using level 2 transistor models relating to a 1μm CMOS process. In these models, threshold voltages of *NMOS* and *PMOS* transistors were high and equal to V_T=*0.75V* and V_T=*-1V*, respectively. Bias and supply voltages were: V_{bias}=*-0.7V*, V_{DD}=*1.5V*, V_{SS}=*-1.5V*, i.e. V_{DD}-V_{SS}=*3V*. Channel length of each transistor was L=1μm. The designed aspect ratios W/L of the transistors are shown in Table 1.

Table 1. Transistor aspect ratios for the circuit of Figure 1

Tr.	M1	M2	M3	M4	M5	M6	M7	M8	M9	M10	M11	M12
W/L	4	4	12	10	4	3	16	16	24	75	30	35

Results of DC simulations are illustrated in Figure 3. The bottom part presents output current I_{OUT} versus the input one I_{IN} for four different values of the tuning voltage, i.e. for V_{TUN}=-0.716V, V_{TUN}=-0.642V, V_{TUN}=-0.568V V_{TUN}=-0.494V. The upper part presents the derivative dI_{OUT}/dI_{IN} versus I_{IN}. As can be seen from the upper diagram, the current amplifier gain A is tuned from -1 to 1.

Linearity of the simulated relation $I_{OUT} = f(I_{IN})$ is not very good. There are two main sources of nonlinearities in the amplifier shown in Figure 1. The first is due to the tuneabilitity, which has been theoretically predicted (see Figure 2), and the second follows from non-linear properties of the current mirrors applied. In the next section, a more linearly operating version of the amplifier is presented, where novel cascode mirrors are used.

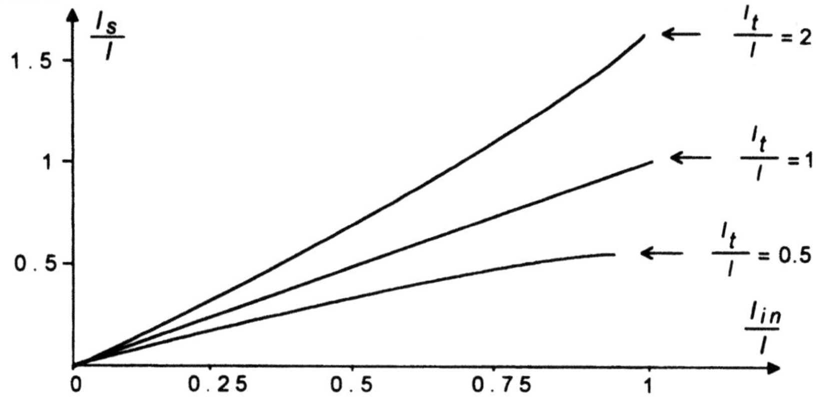

Figure 2. I_S-component of the output current versus I_{IN} (normalized with respect to I)

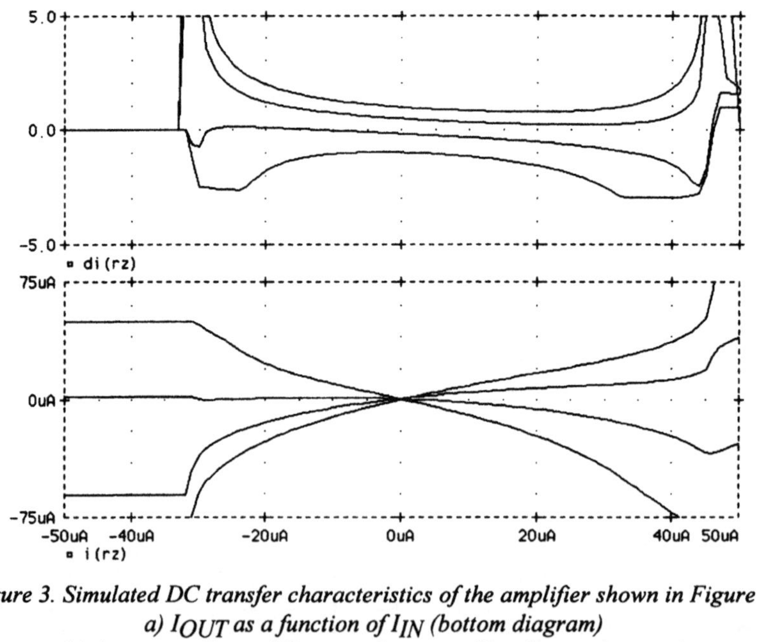

Figure 3. Simulated DC transfer characteristics of the amplifier shown in Figure 1:
a) I_{OUT} as a function of I_{IN} (bottom diagram)
b) derivative dI_{OUT}/dI_{IN} as a function of I_{IN} (upper diagram)

4. CASCODE-BASED REALIZATION

Several cascode-type current mirrors have been published in the literature [3-6]. A very good linearity exhibits the active-input regulated cascode proposed in [5]. This cascode, however, is relatively complex (includes two differential amplifiers of high gain). The cascodes presented in [3] and [4] are simpler but less linear. A half way between the cascodes [3], [4] and that of [5] is the cascode proposed in [6]. Here, a new proposal of the cascode has been applied. Being simpler, it is characterized by a similar linearity as the cascode of [6], i.e. better linearity than that of [3], [4].

A modified version of the current amplifier of Figure 1, with two cascode mirrors of the new type, is shown in Figure 4. One cascode is built of the transistors M5, M9, M13, M14, M15, M16, M17 and the other of the transistors M7, M8, M18, M19, M20, M21, M22. Consider the latter cascode. The concept of achieving a good linearity of its transfer function is to make the transistors M7 and M8 operate with the same changes of their drain-source voltages. Then, the channel length modulation equally affects the transistors M7 and M8. To achieve this, drain current of M19 should be much greater than drain current of M3. Simulation studies of the amplifier shown in Figure 4 were performed for the same conditions as previously. The designed transistor aspect ratios are shown in Table 2.

Table 2. Transistor aspect ratios for the circuit of Figure 4

Tr.	M1	M2	M3	M4	M5	M6	M7	M8	M9	M10	M11
W/L	7	8	24	20	6	6	16	16	34	57	30
Tr.	**M12**	**M13**	**M14**	**M15**	**M16**	**M17**	**M18**	**M19**	**M20**	**M21**	**M22**
W/L	35	32	36	1	20	8	36	20	36	1	8

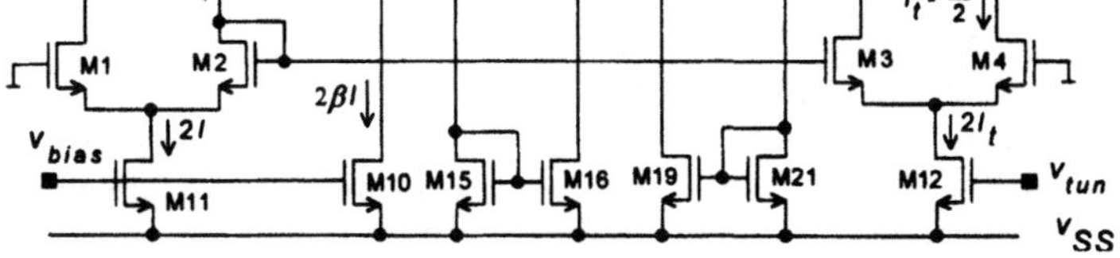

Figure 4. A more linear version of the proposed current amplifier

Results of DC simulations of the amplifier of Figure 4 are illustrated in Figure 5. Comparing Figures 3 and 5 it is seen that the amplifier of Figure 4 exhibits a better linearity of its transfer function than that of Figure 1. This is seen notably from the curves presenting the derivative dI_{OUT}/dI_{IN} (upper diagrams). The allowed range in which the input current I_{IN} can be changed is also wider in the case of the cascode-based amplifier.

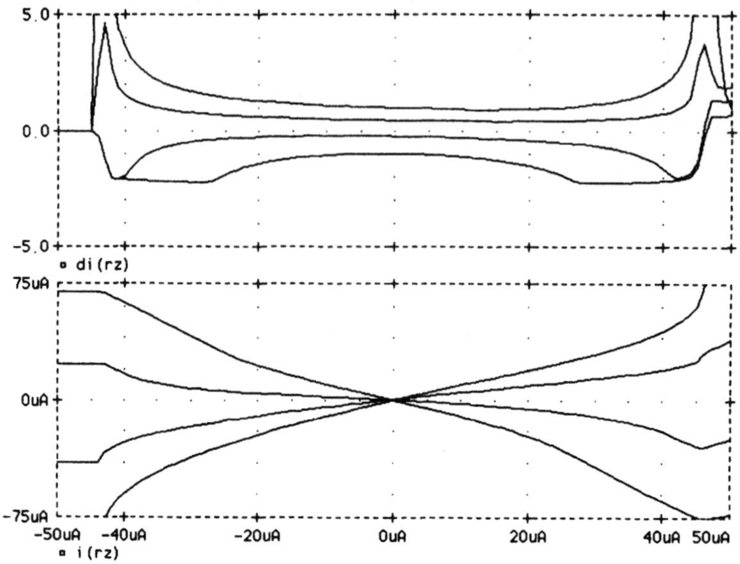

Figure 5. Simulated DC transfer characteristics of the amplifier shown in Figure 4:
a) output current I_{OUT} as a function of input current I_{IN} (bottom diagram)
b) derivative dI_{OUT}/dI_{IN} as a function of I_{IN} (upper diagram)

5. CONCLUSIONS

A novel, operating in class A, current-mode CMOS amplifier has been presented, which is well suited for very low supply voltages. For relatively high absolute values of the transistor threshold voltages, i.e. for V_T=*-1V* (PMOS) and V_T=*0.75V* (NMOS), a proper operation of the amplifier has been achieved with the supply voltages as low as V_{DD}-V_{SS}=*3V*. The advantage of the proposed amplifier is a possibility to realize, according to the controlling voltage V_{TUN}, both positive and negative gains at a single output. This feature makes it attractive to be used in current-mode artificial neural networks to weighting synaptic connections in a programmable way. Two versions of the amplifier have been presented. The first version (Figure 1) is simple while the second (Figure 4), including novel current mirrors, offers a better linearity of the amplifier transfer function.

REFERENCES

[1] Wawryn and B. Strzeszewski, *Low power VLSI neuron cells for artificial neuron networks*, Proceedings of ISCAS'96, Atlanta 1996

[2] Rodriguez-Vazquez, S. Espejo, R. Dominguez-Castro, J. L. Huertas and Sanches-Sinencio, *Current-mode techniques for the implementation of continuous and discrete-time cellular neural networks*, IEEE Transactions on CAS, CAS-40, pp. 132-146, 1993

[3] Sackinger and W. Guggenbuhl, *A high-swing high-impedance MOS cascode*, IEEE Journal of Solid State Circuits, Vol. 25, pp. 289-298, March 1990

[4] Guziński and T. Kulej, *Novel current mirror realization for CCII applications*, Proc. XVth National Conference TOiUE, pp. 78-83, Szczyrk 1992

[5] Serrano and B. Linares-Barranco, *The active-input regulated cascode current mirror*, IEEE Trans. Circuits and Systems-I: fundamental theory and application, Vol. 41, pp. 464-467, June 1994

[6] Palmisano, G. Palumbo and S. Pennisi, *High linearity CMOS current output stage*, Electronics Letters, Vol. 31, pp. 789-790, May 1995

PART II

Power Devices and Thermal Aspects

9

APPLICATION OF INVERSE PROBLEMS TO IC TEMPERATURE ESTIMATION

Marcin Janicki, Mariusz Zubert, Wojciech Wójciak, Mariusz Orlikowski and Andrzej Napieralski

Department of Microelectronics and Computer Science
Technical University of Łódź
al. Politechniki 11, 93-590 Łódź
POLAND

ABSTRACT

In this chapter a method for IC temperature monitoring and estimation will be discussed. First, a brief mathematical description of the problem is presented. The work is focused on finding a solution of heat conduction equation for a thin silicon slab which would be suitable for solving inverse heat conduction problems (IHCPs) consisting in estimating surface heat flux in power ICs. Later, for a given structure, the solution proposed in this paper is compared with the one obtained using finite differences method (FDM).

Next, a short introduction to IHCPs is given. In the presented method, the density of power dissipated in heat sources is estimated solving an inverse problem. For the calculations, the data read from temperature sensors and information about heat sources position are given.

Finally, the influence of input data errors on estimation results is investigated. In order to diminish the influence of these errors, the information from additional temperature sensors is applied to the least squares method (LSM). Additionally, the temperature sensors position influence is also investigated.

1. INTRODUCTION

In the recent few years, due to technological advances, the dimensions used for the design of modern power devices (λ) have become significantly smaller. Thus, the dissipated power density has increased. On the other hand, the total area occupied by power devices is continuously growing. Therefore, the total power dissipation is rapidly increasing. From the above reason the need for thermal analysis methods and tools has arisen.

The power circuit thermal analysis is necessary already in the early stages of the design process. It can help to find the optimal position of heat sources (e.g. power transistors) and temperature sensors as well as the proper radiator dimensions.

Another field where the thermal analysis is required is circuit overheat protection in real working conditions. One of the methods for such overheat protection consists in placing on the circuit a set of temperature sensors. The sensors serve to monitor continuously the temperature and the information from them can be used by control logic unit to commence the system shutdown procedure in case of any malfunction. Up to now, usually, the sensors have to be placed close to heat sources. This solution, however, cannot be applied when the area where temperature sensors can be placed is restricted. Additionally, the placement of temperature sensors close to heat sources could cause interference in the monitored circuit.

This chapter will present a method of thermal IC analysis based on the solution of IHCP. In this method the heat conduction equation has been solved assuming the boundary conditions which are encountered in real integrated power circuits. The detailed mathematical description of the method is presented in the following section.

2. MATHEMATICAL DESCRIPTION

The power integrated circuits are manufactured in a thin silicon slab, where the heat sources are placed at the top and the generated heat is removed at the bottom. For such a slab, the steady state heat conduction process is governed by the following Laplace differential equation [1, 3]:

$$\frac{\partial^2 T}{\partial x^2}+\frac{\partial^2 T}{\partial y^2}+\frac{\partial^2 T}{\partial z^2}=0 \tag{1}$$

where:

T - temperature, *x, y, z - co-ordinates*

In case of power integrated circuits, the boundary conditions necessary for solving the equation (1) can be described as follows (see Figure 1):

- adiabatic surfaces in both horizontal directions (because heat flows mainly towards the bottom of the slab)
- uniform heat removal at the bottom surface modeled by heat exchange coefficient h $[W / (m^2 * K)]$
- power dissipated in heat sources expressed by heat flux coming into the slab

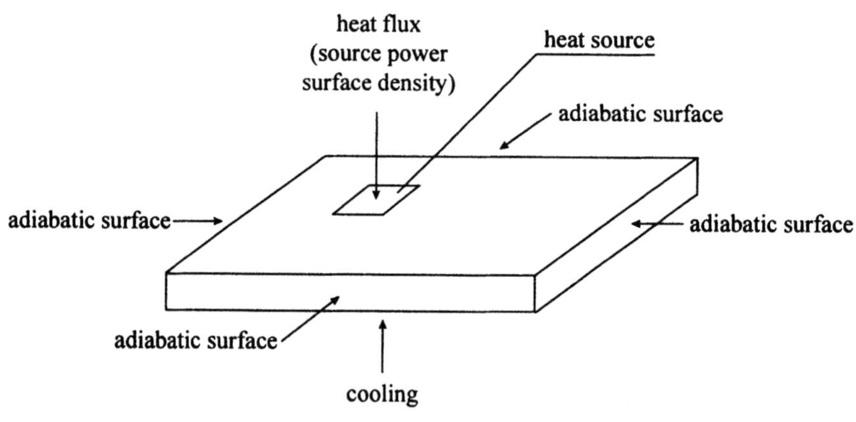

Figure 1. Problem representation

Because the Fourier method renders possible to obtain the solution as a product of functions each dependent only on one coordinate, it has been chosen for the solution of equation (1). Then, for a single square heat source, assuming its constant power density, the solution can be written in the following general form:

$$T(x,y,z)=\sum_{m=1}^{\infty}\sum_{n=1}^{\infty}\left(A_m*\sin(\lambda_m*x)+B_m*\cos(\lambda_m*x)\right)*\left(C_n*\sin(\lambda_n*y)+D_n*\cos(\lambda_n*y)\right)*$$
$$*\left(E_{mn}*\exp\left(-\sqrt{\lambda_m^2+\lambda_n^2}*z\right)+F_{mn}*\exp\left(\sqrt{\lambda_m^2+\lambda_n^2}*z\right)\right) \quad (2)$$

where:

A_m, B_m, C_n, D_n, E_{mn}, F_{mn} - *coefficients;* *m, n - series index;* λ_m, λ_n - *eigenvalues*

Basing on [3], the solution of equation (1) with the earlier described boundary conditions has been obtained in the following form:

$$T(x,y,z)=(A+B+C+D)*q(x,y,z) \quad (3)$$

where:

A, B, C, D - coefficients *q - surface heat flux*

This solution consists of the sum of three infinite series and a certain value A independent of heat sources and temperature sensors positions. It should be mentioned that, for a single heat source occupying the whole slab surface, the equation (3) reduces to the expression $T = [(d / \lambda) + (1 / h)] * q$ which is the so called one dimensional thermal resistance multiplied by heat flux q; where: d - silicon slab thickness, λ - thermal conductivity.

The main advantage of this solution is that the surface heat flux is an explicit factor in the product, therefore it is suitable for estimation purposes. Theoretically, the series in the solutions are infinite, but, because their components are rapidly convergent to zero, they can be truncated.

3. METHODS COMPARISON

In order to verify the correctness of the earlier presented solution, a thin silicon slab with three square heat sources at the top has been considered. The heat sources positions, dimensions and power dissipated in each of them as well as the computed temperature rise at their centres are given in the Table 1. The slab dimensions were 10mm x 10mm x 0.5mm. The heat exchange coefficient value at the slab bottom was equal to 0.1 W / [(mm) ^{2} * K].

Table 1. Information on heat sources

Heat Source	Source center position [mm x mm]	Source size [mm x mm]	Dissipated power [W]	Source temperature rise [K]
A	1.25 x 5.10	0.5 x 0.2	2	21.4
B	6.20 x 2.10	0.4 x 0.2	4	48.7
C	6.30 x 5.10	0.2 x 0.2	3	53.5

The problem has been solved using two different methods. The first one was the method proposed by the authors, the other one was finite differences method (FDM). Then, the surface temperature values, obtained from the two methods, have been compared by subtracting the values taken from the FDM from the values of the analytical solution. The comparison results are presented in Figure 2.

As it can be seen from the Figure 2, both methods give similar results, the maximal temperature rise difference between them is around 10 %. It should be mentioned that the obtained temperature gradient is smaller in case of the FDM method, it is caused by the finite number of mesh nodes, which leads to averaging of temperature values. As the consequence, in the FDM results the maximal temperature rise is underestimated. Additionally, the effects of structure discretisation in FDM can be observed.

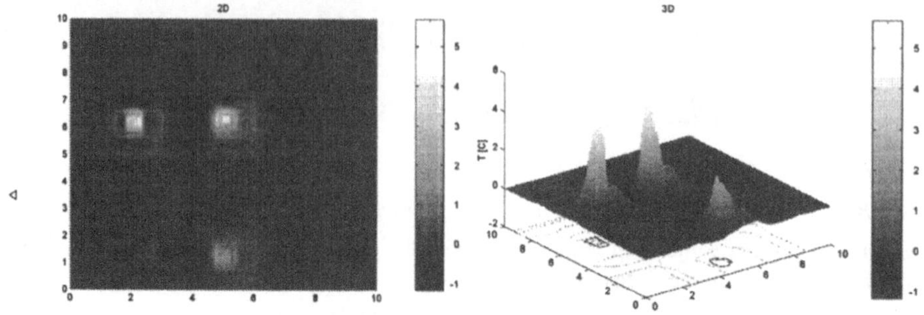

Figure 2. Comparison results

4. APPLICATION OF INVERSE HEAT CONDUCTION PROBLEMS TO POWER AND TEMPERATURE ESTIMATION

The temperature field can be calculated solving a direct problem, if all the coefficients in the equation describing heat conduction as well as boundary and initial (for unsteady states) conditions are known. Unfortunately, it often happens that some data required for solving this equation are not known, then, an inverse problem has to be solved. Considering the goal of solving inverse problems, they usually consist in determining the following parameters [2]:

- boundary conditions
- initial conditions
- position of heat sources
- power dissipated in heat sources
- coefficients describing material thermal properties

This work will be focused on the estimation of dissipated power. The problem consists in estimating the dissipated power and temperature in a structure. In this case, the whole steady state temperature map is calculated knowing the heat sources positions and temperature values only at selected points of structure. For the estimation purposes, the mathematical solution presented in the previous section has been used. Then, invoking the problem linearity, for n heat sources and m temperature sensors, the equation (3) can be written in the following matrix form:

$$T = A \cdot Q \tag{4}$$

where:

$T = \begin{bmatrix} T_1 & T_2 & \cdots & T_m \end{bmatrix}^T$ - *sensor temperatures*

$Q = [q_1 \ \cdots \ q_n]^T$ - *unknown power densities*

$A = \begin{bmatrix} a_{11} & \cdots & a_{1n} \\ \cdots & \cdots & \cdots \\ a_{m1} & \cdots & a_{mn} \end{bmatrix}$ - *coefficients matrix*

If the temperature sensors positions are chosen so that matrix A determinant is not equal to 0, the density of power dissipated in each of the heat sources can be estimated using the following equation:

$$Q = A^{-1} \cdot T \tag{5}$$

As it can be seen, the minimal required number of temperature sensors m is equal to the number of heat sources. However, as it will be shown in the next section, additional temperature sensors may be placed in order to minimize the error introduced to sensors temperature measurements.

5. ERROR CORRECTION

Because inverse problem solutions are usually sensitive to input data errors, the possibility of error correction is discussed in this section. In order to illustrate this problem, a simple experiment has been conducted. For the experiment, the same structure as for the method comparison has been used. The temperature values have been read in the four corners of the slab top surface at 1 mm from the edges. Then, for all the possible combinations of three required sensors (because there are three heat sources), the dissipated power has been estimated solving inverse problems. The estimation results are very accurate in case of errorless input data. However, in case when there are some errors introduced to these data, the power estimation error becomes quite significant (see Table 2). As it has been shown in [5], it is the condition number of matrix A in equation (4) relating input and output relative errors, which decides about the sensitivity to input data errors. The condition number of matrix A is equal to $||A|| * ||A^{-1}||$, where $||A||$ is a norm of matrix A and A^{-1} is an inverse matrix to A. In the best case the condition number was equal to 11.49. For the worst case (condition number 391.32) the estimation results have become senseless in the presence of input data errors.

One of many possible ways to minimize the error influence on the inverse problem solution is to place on the circuit additional temperature sensors. The power estimation error can be then lessened using redundant information from these sensors. The most commonly used method for error minimization is the least squares method (LSM) [4]. This method consists in finding such an estimate of vector Q in equation (5), which would minimize the following norm:

$$||A * q - T||^2 \tag{6}$$

The optimal estimate of vector Q, for which the norm derivative is equal to 0, is given by:

$$Q = (A^T * A)^{-1} * A^T * T \tag{7}$$

The above equation was implemented in order to estimate the power dissipated in heat sources. The number of sensors has been increased up to five (four in the corners and one in the middle), what has made the condition number of matrix A drop down to 5.19. The experiment results are shown in Table 2.

Table 2. Estimation results

		Source	Number of sensors		
			3 worst case	3 best case	5
errorless input data	Estimated power [W]	A	2.00	2.00	2.00
		B	4.00	4.00	4.00
		C	3.00	3.00	3.00
	Estimation error [%]	A	0.00	0.00	0.00
		B	0.00	0.00	0.00
		C	0.13	0.00	0.00
error module up to 5 % mean error value equal to 0	Estimated power [W]	A	1.62	2.05	1.96
		B	2.94	3.10	4.10
		C	14.24	3.12	2.99
	Estimation error [%]	A	19.00	2.50	2.00
		B	26.50	22.50	2.50
		C	474.67	4.00	0.33

As it can be seen from Table 2, the proposed method is very accurate in case of errorless data. However, the method proved to be quite sensitive to the input errors introduced by temperature sensors. Even slight distortions of input data may cause quite significant errors, the worst case in the example was 475% error of temperature estimation for the input data distorted only up to 5%. As it can be seen, the error can be diminished using the LSM, which takes advantage of the information from redundant sensors. In the worst case mentioned earlier, the error has been brought down almost to zero value.

6. CONCLUSIONS

A new method for power and temperature estimation in power ICs has been presented. The possibility of estimating the dissipated power and temperature distribution all over the monitored structure by solving thermal inverse problem has been shown. The main advantage of such a solution is that temperature can be, at least theoretically, measured at any point of the structure. Temperature sensors need not be placed close to heat sources, so the monitoring circuit does not interfere with the power circuit.

The problem has been posed in such a way that it has been possible to obtain the solution of the heat conduction equation which can be applied for integrated circuits thermal analyses. Because the surface heat flux is an explicit factor in the problem solution, it is possible to use the solution for the dissipated power estimation solving an inverse problem.

In comparison to the finite differences method, the hereby proposed one gives very accurate results which do not depend on number and position of mesh nodes, what occurs while using the FDM or other numerical methods. Additionally, for a given structure, all the coefficients relating sensor positions to surface heat sources need to be calculated only once.

The experiment has showed that the main disadvantage of the presented method is its sensitivity to input data errors. One of possible ways to reduce the errors influence, for a given heat sources configuration, is to optimize the temperature sensors positions in order to obtain minimal sensitivity to input data by the minimization of the condition number of coefficient matrix A. Another possible solution is to decrease the errors influence by placing in the circuit redundant temperature sensors. The information from these sensors can be used for the reduction of errors influence using one of the well-known mathematical methods, e.g. least squares method or minimal energy method (see [4]). The hereby described method will be the subject for the further research.

REFERENCES

[1] S. Wiśniewski, *Wymiana ciepła*, PWN, Warszawa, 1988 (in Polish)

[2] W. Szargut, *Modelowanie numeryczne pól temperatury*, WNT, Warszawa, 1992 (in Polish)

[3] E.Kącki *Termokinetyka*, PWN Warszawa 1967 (in Polish)

[4] D. Lesnic, L. Elliott and D. B. Ingham, *Application of the boundary element method to inverse heat conduction problems*, Int. J. Heat Mass Transfer, Vol.39, No.7, pp. 1503-1517, 1996

[5] G. Dahlquist and A. Bjoerck, *Metody numeryczne*, PWN Warszawa 1983 (in Polish)

10

THERMAL MODEL FOR MCM'S

Francesc Masana

GDS-DEE, UPC
C/J. Girona 1-3, C-4, 08034 Barcelona
SPAIN

ABSTRACT

A model based on closed form expressions for steady state thermal resistance evaluation of chips and their thermal coupling, under adiabatic and isothermal boundary conditions, is presented.

The expressions for thermal resistance use the variable heat spreading angle approximation and the thermal coupling is calculated using the Green's function approximation and the method of images.

The model shows good accuracy and behaviour over the conditions most often encountered in practice, so it can be used in many design situations.

1. INTRODUCTION

Thermal analysis of multiple device structures, as MCM, needs to consider both self-heating of devices and thermal coupling between them. This problem has been traditionally approached using analytical methods [1,2] or numerical techniques [3]. In both cases, the arrived solution embodies the mutual effects in problems including more than one heat source.

There are many situations, however, where the complexity of such methods precludes their use until some degree of concretion in the design is attained. In such cases, models based on the thermal resistance concept [4-6] can be useful. However, if thermal coupling stands for an important part of temperature rise on some or all of the devices, a way of considering the mutual effects has to be included.

In what follows, a closed form method for self-heating evaluation through the thermal resistance concept, jointly with the calculation of cross coupling effects between devices is presented.

2. THE MODEL

The proposed model is sketched in Figure 1, where the thermal-electrical duality is used.

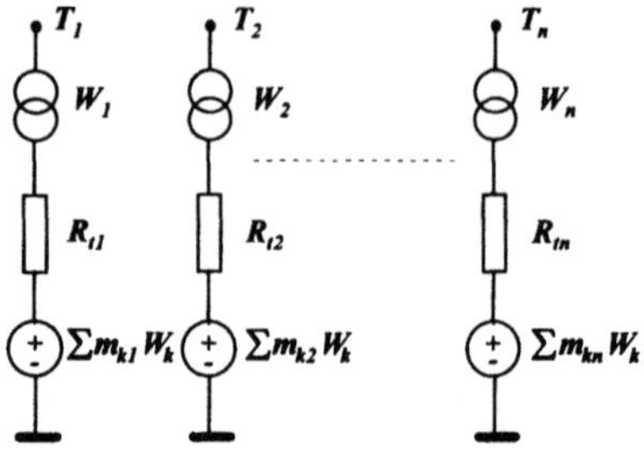

Figure 1. Model representation

The power dissipated in each element is modelled through a current generator of value W_i, the self heating through the element's thermal resistance R_{ti} and the thermal coupling by means of current dependent voltage generators with mutual resistance m_{ki}. Then, the temperature of each individual element can be calculated as:

$$T_i = R_{ti} \cdot W_i + \sum_{j \neq i} m_{ji} \cdot W_j \tag{1}$$

The model parameters are the thermal resistance R_{ti} of each element and the mutual thermal resistances m_{ki} from every other element to it.

3. THE EVALUATION OF THERMAL RESISTANCE

It has been often assumed that heat spreads with a constant angle from the power dissipating element [1], 45° being the most commonly used. The value for the thermal resistance of a square element of side $2l$ on a semi-infinite substrate of thickness w and thermal conductivity k is:

$$R_t = \frac{1}{4 \cdot k \cdot l} \cdot \frac{w}{l + w \cdot tan\alpha} \tag{2}$$

For our development with a finite substrate, we define the dimensionless spreading resistance coefficient and system geometry parameters as:

$$H = 4 \cdot k \cdot l \cdot R_t ; \quad l_n = l/L, 0 \leq l_n \leq 1; \quad w_n = w/L, 0 \leq w_n \leq \infty \tag{3}$$

where l and L are the (square) element and substrate half side.

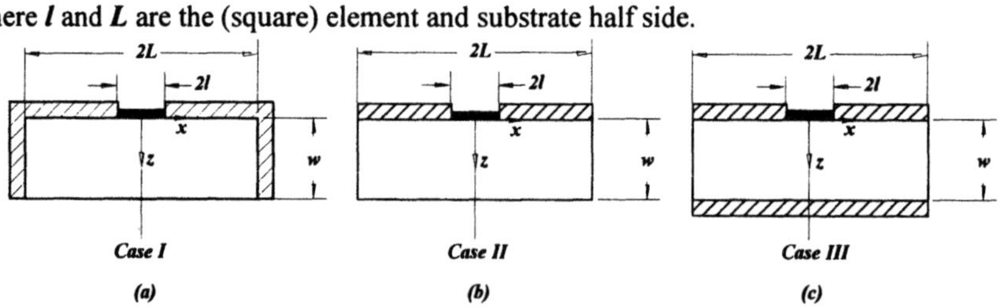

Figure 2. Boundary conditions: a) Case I, b) Case II and c) Case III

The variable heat spreading angle method, described in detail elsewhere [5,6] will now be applied to find the thermal resistance for the following sets of boundary conditions, sketched in Figure 2, where the top surface is considered adiabatic in all cases:

- *Case I*: Side walls adiabatic and bottom surface isothermal.
- *Case II*: Side walls and bottom isothermal at the same temperature.
- *Case III*: Side walls isothermal and bottom surface adiabatic.

3.1. CASE I

The considerations made for establishing a relationship between the spreading angle and the Case I boundary conditions are [5]:

- For $l_n << 1$ is $\alpha \approx 45°$, because sidewall does not constrict the flow.
- For $l_n \rightarrow 1$, $\alpha \rightarrow 0$, due to sidewall (adiabatic) constriction of flow.
- For $w_n/l_n << 1$, $\alpha \approx 0$, due to the close proximity of bottom surface.

They lead to an expression for $tan\alpha$:

$$(tan\alpha)_I = (1 - l_n) \cdot \frac{w_n}{l_n + w_n} \tag{4}$$

For substrates where $w_n > 1$, however, the use of expression (4) together with expression (2) will imply a spread beyond the substrate boundaries, which is physically impossible. To overcome this, a truncation is introduced as depicted in Figure 3, leading to an expression for thermal resistance:

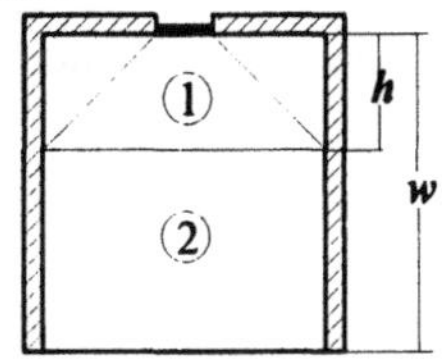

Figure 3. Truncation for thick substrates

$$R_t = \frac{1}{4 \cdot k \cdot l} \cdot \left(\frac{h}{l + h \cdot tan\alpha} + \frac{w - h}{L^2} \cdot l \right) \tag{5}$$

Putting this all together, gives the following expression for H:

$$H_I = \frac{h_n}{l_n + h_n \frac{w_n}{w_n + h_n} (1 - l_n)} + (w_n - h_n) l_n \tag{6}$$

where h_n is defined as [5]:

$$h_n = tanh(w_n) \tag{7}$$

3.2. CASE II

The sidewall and bottom surfaces are now isothermal and at the same temperature, so they will not constrict the flow at all. This means that the amount of heat flowing through the sidewall increases as l_n approaches unity, in contrast to what we had in Case I. A sketch to illustrate the situation is depicted in Figure 4.

Truly bidimensional situations are not easily ammenable to closed form so, being H a purely geometrical and dimensionless parameter, we have to find a pseudo-bidimensional equivalence to convert the lateral flow labelled ① in Figure 4 into an equivalent vertical flow ②. Summarizing, we have:

- The dependence of H on w_n is the same as for Case I.
- Because now the heat spreading is not constricted by the sidewall, we use w_n in place of h_n and drop out the second term in (6).

- The spreading angle will now increase with l_n.
- For l_n much smaller than unity, the problem reduces to Case I.

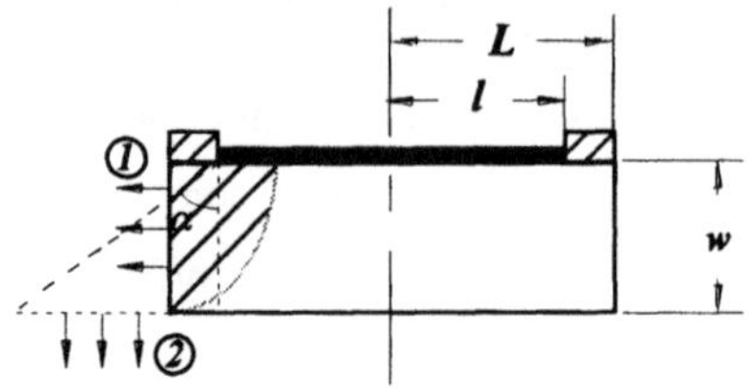

Figure 4. Flow through the sidewall. Case II

We can thus write:

$$H_{II} = \frac{w_n}{l_n + w_n \dfrac{w_n}{l_n + w_n}\left(1 + n\cdot l_n\right)} \tag{8}$$

where n is a parameter to adjust the maximum value for α. A good fit is obtained with $n = 0.732$ ($\alpha_{max} = 60°$) [6].

3.3. CASE III

Now the heat leaving our system can flow through the sidewalls only. The problem is purely bidimensional and so a different approach is needed.

If we consider that for $w_n << l_n$ the heat flows laterally for nearly all of its path, while for $w_n >> l_n$ the path is almost vertical, it will be possible to divide the flow into two parts, as shown in Figure 6*a*: one vertical under the heating element, labelled as ①, and one horizontal away from it, labelled as ②. Both parts have been calculated in [6] under the following additional assumptions:

- For $w_n \to \infty$, the bottom surface has no influence on the spreading, irrespectively of the imposed condition.
- For $w_n \to 0$, the flow tends to be purely horizontal, meaning $\alpha = 90°$.

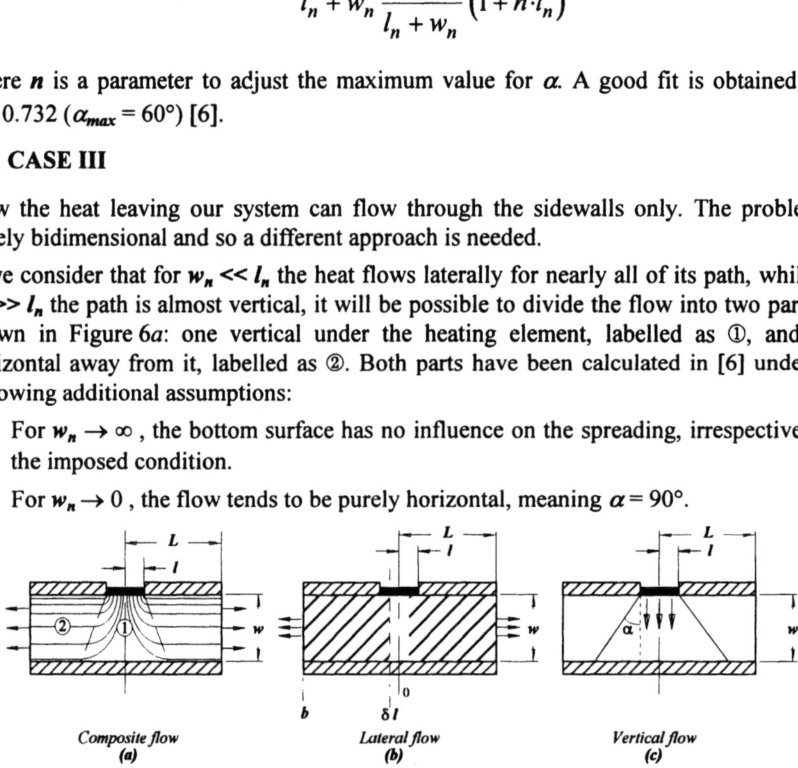

Figure 6. Sketch of flow lines (a) illustrating its decomposition in two parts: b) Lateral flow, c) Vertical flow

The final value of the spreading coefficient is given in the following expression, where $\delta(= exp(-1/2))$ is adjusted for $w_n << 1$ and $l_n \approx 1$.

$$H_{III} = \frac{l_n}{2w_n}\cdot ln\frac{1}{\delta l_n} + \frac{w_n}{l_n + w_n \dfrac{l_n + w_n}{w_n}\left(1 + n\cdot l_n\right)} \tag{9}$$

4. EXTENSION TO RECTANGULAR GEOMETRY.

The extension of the model to a rectangular substrate of dimensions $2L_1$x$2L_2$ with a heat source of dimensions $2l_1$x$2l_2$, is illustrated in [5] for Case I conditions. It requires the calculation of the integral:

$$R_t = \frac{1}{4k}\int_0^w \frac{dz}{(l_1 + z\cdot tan\alpha)\cdot(l_2 + z\cdot tan\beta)} \tag{10}$$

where α and β are two different spreading angles, one for each substrate dimension. In order to simplify the final expression, a new set of variables is defined as follows:

$$\frac{l_2}{l_1} = \gamma_e \;;\; \frac{L_2}{L_1} = \gamma_s \;;\; l_{1n} = \frac{l_1}{L_1} \;;\; w_n = \frac{w}{L_1} \;;\; H = 4\cdot k\cdot l_1 \cdot R_t \tag{11}$$

The spreading angles can be then written as:

$$tan\alpha = (1 - l_{1n})\cdot\frac{w_n}{w_n + l_{1n}} \;;\; tan\beta = \left(1 - l_{1n}\cdot\frac{\gamma_e}{\gamma_s}\right)\cdot\frac{w_n}{w_n + l_{1n}\cdot\gamma_e} \tag{12}$$

and the spreading resistance factor H is:

$$H_I = \frac{1}{\gamma_e\cdot tan\alpha - tan\beta}\cdot\ln\frac{l_{1n} + h_n\cdot tan\alpha}{l_{1n} + h_n\cdot tan\beta/\gamma_e} + (w_n - h_n)\cdot\frac{l_{1n}}{\gamma_s} \tag{13}$$

Equation (13) is only valid if γ_e and γ_s are not both unity.

5. ASYMMETRICALLY LOCATED HEAT SOURCE.

If the heat source is not located symmetrically onto the substrate, a way to extend the model is by superposition of two symmetrical cases [5].

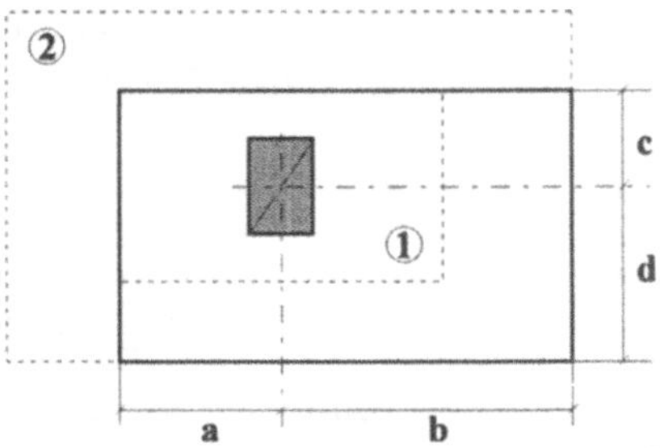

Figure 7. Illustration of superposition method for the asymmetric case

The approach used is shown in Figure 7. The overall thermal resistance R_t is calculated as the parallel combination of two resistances, R_{t1} and R_{t2} given by: $R_{t1} = 2\cdot f(L_1 = 2a, L_2 = 2c)$ and $R_{t2} = 2\cdot f(L_1 = 2b, L_2 = 2d)$, where f is given by (6) (8) or (9).

6. THE THERMAL COUPLING

The exact expression for the radial temperature distribution around a circular element of radius r_0 dissipating a constant power per unit area in the surface $z = 0$ of a semi-infinite substrate has been calculated [7] as a function of the complete elliptic integrals of first and second kind. Expanding the elliptic integrals in power series gives:

$$T(0,r) = \frac{W}{2\pi kr}\left[1 + \frac{1}{8}\left(\frac{r_0}{r}\right)^2 + \cdots\right] \quad for \;\; r > r_0 \tag{15}$$

where the first term is the Green's function for the temperature at a point ***(0,r)*** due to a total power ***W*** concentrated as a delta function at the origin, in the half-space $z > 0$ (or $z < 0$ for the sake).

In the study of thermal coupling we seek only for the temperature outside the heating element, so the maximum error incurred by dropping the quadratic term will be 12%, which is considered acceptable.

If we now substitute ***T(0,0)*** calculated from the exact solution in [7] into (15) and drop all the terms but the first one, we obtain:

$$T(0,r) \cong T(0,0) \cdot \frac{r_0}{2r} \cong W \cdot R_t \cdot \frac{r_0}{2r} \tag{16}$$

where ***T(0,0)*** is calculated as the product of the element power, times its thermal resistance. Substituting (16) into (1) we get:

$$T_i = R_{ti} \cdot W_i + \sum_{j \neq i} R_{tj} \cdot \frac{r_{0j}}{2r_{ji}} \cdot W_j \tag{17}$$

Most of the applications found in practice, however, deal with square or rectangular elements. For the square element, we will take one half of the side length as the radius r_0. For the rectangular one, the approach is developed in [8], where a line instead of a point source is used. The errors incurred by those approximations will be larger for points close to the source. However, the minimum distance between the center points of two equally sized elements is twice their radii or their half side, so the error in most practical cases will still be acceptable.

If now the substrate dimensions are finite, we can substitute the Neumann or Dirichlet boundary conditions by mirror image heat sources or sinks [2] as shown in Figure 8 for Case I conditions.

Only the first level of images is shown in the figure. In fact, the new images will have their own images as well, up to an infinite number of them. Fortunately, the successive layers of images will be much farther from the point of interest (located within the original substrate) and the error incurred neglecting them is getting smaller as we move away.

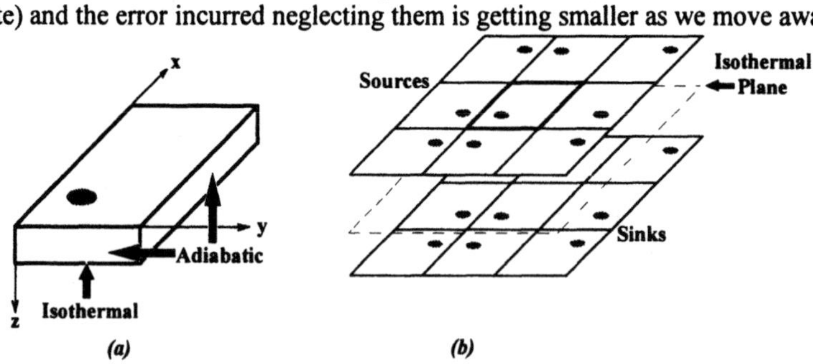

Figure 8. Method of images. a) Actual substrate. b) First level images

The temperature at any point due to a single heat source will be:

$$T(0,r) = T(0,0) \cdot \sum_i \frac{r_0}{2r_i} \tag{18}$$

where i extends to the heat souce and its images, and the sign is into r_i.

7. RESULTS AND CONCLUSION

To emphasize both the self-heating and the thermal coupling aspects of the model, we will use a structure consisting of a GaAs FET on a MMIC.

Table 1. Comparison of model against numerical results from [3].
Values displayed are Temperature increments above the isothermal chip bottom

FINGER #	1	2	3	4	5	6	7	8
T (num.) [3]	26.0	30.0	32.5	34.2	35.5	36.2	36.8	37.1
T (model)	27.6	30.3	32.0	33.2	34.1	34.6	35.0	35.1
FINGER #	**16**	**15**	**14**	**13**	**12**	**11**	**10**	**9**
T (num.) [3]	26.0	30.0	32.5	34.2	35.5	36.2	36.8	37.1
T (model)	27.7	30.4	32.1	33.3	34.1	34.6	35.0	35.2

Chip size is 1.7x1.1x0.125mm, and the FET is composed of sixteen fingers of 150μmx1μm with 15μm pitch, dissipating a total power of 545mW, its center point being at 375μm and 500μm from the closest chip edges. The results are summarised in Table 1, and show a good agreement with those obtained by numerical simulation in [3].

For the model calculations, Case I boundary conditions have been used and the GaAs thermal conductivity is taken as 0.35 W/cmK. Finger cross coupling stands for between 40 and 55 % of temperature rise.

REFERENCES

[1] R.F. David, *Computerized Thermal Analysis of Hybrid Systems,* IEEE Tr. PHP, PHP-13, No. 3, Sept 1977, pp 283-290.

[2] A.L. Pallisoc and C.C. Lee, *Exact thermal representation of multilayer rectangular structures by infinite plate structures using the method of images,* JAP, VOL 64, NO 12, Dec. 1988, pp 6851-6857.

[3] D.H. Chien, C.C. Lee, M. Rachlin, A. Peake and T. Kole, *Thermal Analysis of Packaged GaAs Devices Using Chip Model with Finite Element Method,* Int. Journal MEP, VOL 20, No 1, 1st Q 1997, pp 3-11.

[4] W. Dobbelaere, L. Matthys, E. De Baetselier, W. Goedertier and G. De Mey, *Heat Spreading Angles in Multilayer Structures,* MIXDES'96, Lodz, Poland, June 1996, pp 238-243.

[5] F.N. Masana, *Closed Form Solution of Junction to Substrate Thermal Resistance in Semiconductor Chips,* IEEE Tr. CPMT, Part A, CPMTA-19, Dec. 1996, pp 539-545.

[6] F.N. Masana, *Closed Form Thermal Resistance Calculation for Different Sets of Boundary Conditions,* MIXDES'97, Poznan, Poland, June 1997, pp 241-246.

[7] J. Frey, *Multimesa Versus Annular Construction for High Average Power in Semiconductor Devices,* IEEE Tr. ED, ED-19, No 8, Aug. 1972, pp 981-985.

[8] F.N. Masana, *Thermal Coupling Between Heat Dissipating Elements in Multiple Device Structures,* MIXDES'97, Poznan, Poland, June 1997, pp 295-300.

11

CONTRIBUTION OF RADIATION IN HEAT DISSIPATION IN ELECTRONIC DEVICES

[1]Bogusław Więcek and [2]Gilbert De Mey

[1] Technical University of Łódź
Institute of Electronics
ul. Stefanowskiego 18/22, 90-924
POLAND

[2] University of Gent
ELIS, Sint Pieterniuwstraat 41, B-9000 Gent
BELGIUM

ABSTRACT

This paper presents numerical and experimental results of heat transfer by radiation, convection and conduction in hybrid microelectronic circuits. We chose a heat source with a non-uniform temperature distribution, which agrees with typical cases frequently met in electronics. In this work we evaluate a complex heat transfer coefficient including the non-linear phenomenon for radiative and convective heat dissipation.

1. INTRODUCTION

It has already been underlined by many authors that in order to evaluate more precisely the temperature in microelectronic devices, one should include convection and radiation into the modelling of heat transfer [1-5]. In many previous works, convection was simply approximated by transfer coefficient taken from the tables and being typically constant over the temperature range. Because of a multidimensional nature of convection and the non-linear characteristics for both the convection and the radiation, until now there are not many works which include these phenomenon into the entire heat removal modelling [1,4]. In this work we model conduction together with radiation and convection, applying their non-linear parameters, and including the temperature-dependent emissivity, e.g. for metals. Simulations yield non-uniform temperature distributions in the electronic device, and using these distributions we evaluate the non-linear complex heat transfer coefficient, comparing it to published and measured data.

2. CONJUGATE MODEL OF HEAT DISSIPATION

In this work we present a numerical model of heat removal for a hybrid long resistor placed on a ceramic substrate (Figure 1). Because $l/w>>1$, where l and w are the length and the width of the resistor, the problem can be reduced to a one-dimensional modelling in heat conduction in the substrate. This is because of the symmetrical heat transfer in vertical direction, neglecting the fact that the upper side of the substrate is cooled down worse, because of warm fluid moving up and developing a convective boundary layer [5].

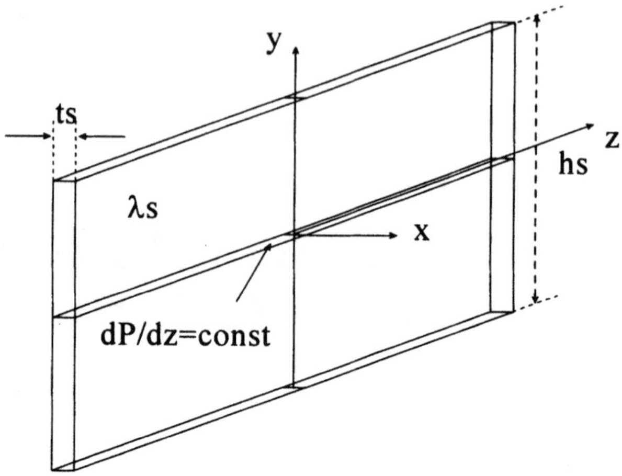

Figure 1. Substrate with the heat source

The model starts from the energy conservation law for the substrate expressed by heat fluxes:

$$-S_s dy \frac{d\varphi_s(y)}{dy} = 2S_a(\varphi_r + \varphi_c) \tag{1}$$

where: φ_s, φ_r, φ_c denote the heat fluxes that correspond to conduction in the substrate, radiation and convection from the source to the ambient. S_s and S_a denote the areas that the fluxes pass through, as shown in Figure 2.

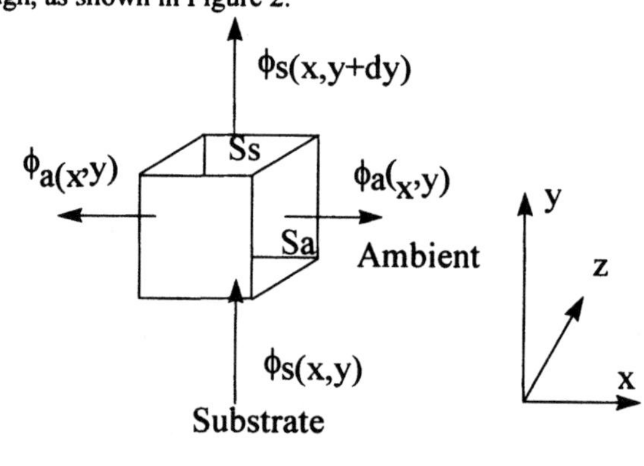

Figure 2. Heat fluxes in the substrate

The fluxes are described as follows:

$$\varphi_s = -\lambda_s \frac{\partial T}{\partial y} \qquad \varphi_c = \alpha_c (T - T_a) \qquad \varphi_r = \varepsilon\sigma(T^4 - T_a^{\ 4}) \tag{2}$$

where: α_c - convective heat transfer coefficient, ε - emissivity, σ - Boltzman constant, λ_s- substrate thermal conductivity and T_a - ambient temperature.

In the model we include temperature dependent local convective heat transfer coefficient α_c in the form of equation (3) [4]. In practice the heat transfer coefficient for non-linear temperature distribution in the substrate does not increase so much as it has been obtained by non-conjugate modelling using analytical and semi-analytical approaches [1-2].

$$\alpha_c = \alpha_0 \left(\frac{\Delta T}{\Delta T_0} \right)^m \tag{3}$$

where m=0.2 for substrates with non-linear temperature distribution.

Replacing heat fluxes in equation (1) using (2) and (3) one can get the non-linear energy equation for sourceless element in the substrate.

$$t_s \lambda_s \frac{d^2 T}{dy^2} = 2\varepsilon\sigma\left(T^4 - T_a^4\right) + \frac{2\alpha_0}{\left(\Delta T_0\right)^m}\left(T - T_a\right)^{1+m} \tag{4}$$

with the following boundary conditions describing the heat source P_z=dP/dz in the middle of the substrate, adiabatic condition on its edge. It is assumed that half of the power is dissipated in each half of the substrate (Figure 1).

$$\left.\frac{\partial T}{\partial y}\right|_{y=0} = \frac{P_z}{2\lambda_s t_s}, \qquad \left.\frac{\partial T}{\partial y}\right|_{y=\pm\frac{h_s}{2}} = 0 \tag{5}$$

In this work it is assumed that heat exchange by radiation is between the microelectronic device with the emissivity ε and the infinite ambient or the black body with its emissivity ε =1 as shown in equation (2).

Equation (4) and (5) can be solved using Runge-Kutta method.

3. EMISSIVITY EVALUATION

The results of the reflection - transmission measurements and the calculated emission are presented in this chapter. We measured both aluminium samples with different surface conditions and semiconductors, i.e. silicon and germanium, using the IR Spectrometer Model 1725 by Perkin-Elmer. All measurements were done in the direction normal to the surface of investigated body, in the spectral range of 1.5÷25μm.

For opaque materials like aluminium we evaluate the normal spectral emissivity as:

$$\varepsilon_n(\lambda) = 1 - \rho_n(\lambda) \tag{6}$$

where $\rho_n(\lambda)$ denotes the normal spectral reflectivity.

For semitransparent semiconductors or diamond-like samples we include transmission of the material as well. It is quite easy to evaluate emissivity for samples large in comparison with the wavelength, where there is no any internal reflections and wave interference effects. In such cases normal emissivity can be expressed as.

$$\varepsilon_n(\lambda) = 1 - \rho_n(\lambda) - \tau_n(\lambda) \tag{7}$$

where $\tau_n(\lambda)$ denotes the normal spectral transmissivity

For multilayer structures we include internal reflection, which obviously takes place in the practice By off-line calculations using the optical constants identified during the measurement, one can find parts of the energy absorbed in every layer, what directly corresponds to the layer's emissivity.

Assuming that we have a partially transmitting layer, the total absorption A including the reflection from the second surface is [1]

$$A = \frac{(1-\rho)(1-\tau)}{1-\rho\tau} \tag{8}$$

Equation (8) was derived by net-radiation method, under the assumption of isothermal conditions (absorption does not increase the temperature). Similarly we can obtain the reflected and transmitted energy fractions, which were measured in this work.

$$R = \rho\frac{1+(1-2\rho)\tau^2}{1-\rho^2\tau^2} \qquad T = \frac{\tau(1-\rho)^2}{1-\rho^2\tau^2} \tag{9}$$

From equation (9) we can evaluate reflectivity ρ and transmissivity τ, as function of wavelength λ. Together with equations (8) and (9), these values can then be used to evaluate the emissivity for any dielectric material with and without internal reflections. Using equation (9) we assume material thickness much larger than the applied wavelength, otherwise we need to consider wave interference, as well. One could notice that A=ε for large samples where there are no internal reflections and wave interference.

Table 1. Radiation properties for semiconductors

	R	T	A	ρ	τ	ε	n
Si 2.5-5.5µm.	0.46	0.53	0.01	0.30	0.29	0.39	3.43
Si 8-12µm.	0.38	0.36	0.26	0.30	0.21	0.48	3.45
Ge 2.5-5.5µm.	0.54	0.46	0.00	0.37	0.37	0.26	4.10
Ge 8-12µm.	0.54	0.45	0.01	0.37	0.37	0.26	4.14

The parameter values obtained agree with published values and measured by other methods, e.g. the refraction indexes from [5] are: $n_{Ge}=3.99$, $n_{Si}=3.49$.

Besides of semiconductor there are various metals used in microelectronics e.g. aluminium. The total (integrated over wavelength), normal emissivity $\varepsilon_{T,n}$ for metals strongly depends upon the temperature, giving the next non-linear phenomenon included in this work [1,2,3].

$$\varepsilon_{T,n} = 0.576\sqrt{\rho(T)T} - 0.124\rho(T)T$$

$$\rho(T) = \rho_{273}\frac{T}{273} \tag{10}$$

where ρ(T) denotes the electrical resistivity,

Applying the total emissivity of metal as $\varepsilon_T \approx 1.2\varepsilon_{T,n}$ [2], the energy equation (4) can be extended using (9), giving the results presented in *Figure 3* and *Table 3.*

There are chosen experimental results presented in Figure 3 and 4. It was already underlined by many authors that the emissivity strongly depends on surface conditions of a body. E.g. non-polished aluminium has very low emissivity, while a thin, of few-µm. thickness of diamond-like layer can significantly increased this value, as shown in Figure 3.

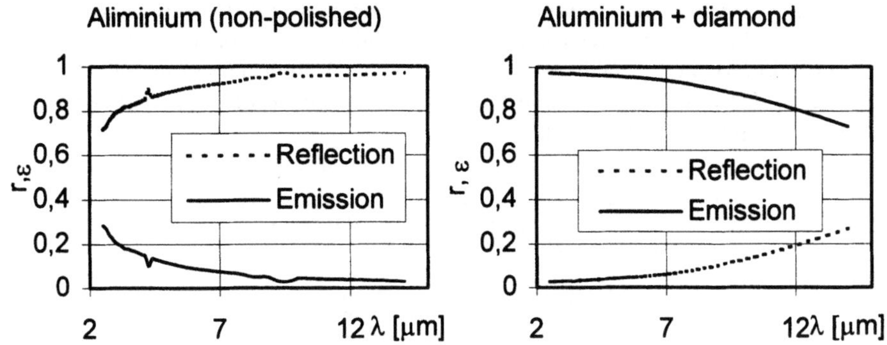

Figure 3. Reflection and emission for aluminium

Semiconductors (Si, Ge) are transparent for infra-red radiation as shown in Figure 4. It is because of small thickness of the semiconductor wafer used as a substrate in microelectronics. The real emissivities obtained for samples without internal reflection are presented in Table 1.

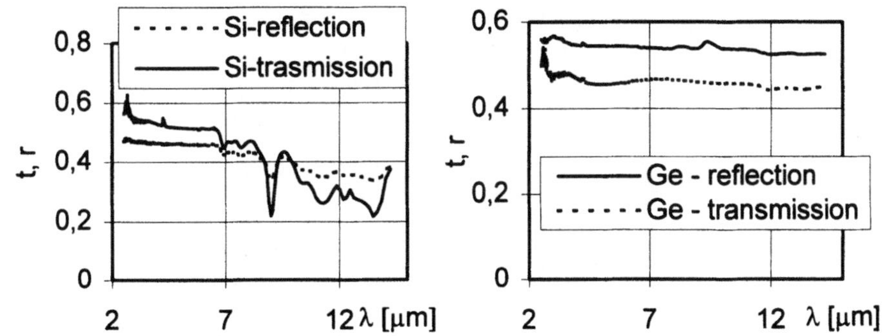

Figure 4. Reflection and transmission for semiconductors

4. RESULTS

The level of the power delivered to the heat source has an influence on the maximum temperature in the substrate. The relation T_{max} *versus power* P_z is non-linear. We found the influence of the power dissipated in the substrate on the temperature and the heat transfer coefficient.

The mean value of heat transfer coefficient α is defined as:

$$\alpha_{T_{mean}} = \frac{P_z}{2h_s\left(T_{mean} - T_a\right)} \tag{11}$$

The simulations was performed for black body *(ε=1)* and for a convective heat transfer coefficient varying according to equation (3) [5]. The parameters of the ceramic substrate and the air are presumed not to depend upon temperature and are as follows: T_a=300K, λ_s=20W/Km, t_s=0.4mm, h_s=3cm. The parameters of the non-linear model of convection according to equation (3) are: α_0=7W/m^2, ΔT_0=20°C

A long hybrid resistor was a heat source placed in the middle of ceramic substrate as shown in Figure 5. Using thermographic system TVS4000-HUGHES, supported with a computer interface for capturing the images and powerful software running under Windows95. For the measurement we used a hybrid resistors circuit, but only the thick one, placed nearly in the middle was powered (Figure 5). Both sides of the substrate were covered by thin coating with emissivity $\varepsilon \approx 0.9$. The measurements were very close to the modelling, as presented in Figure 6.

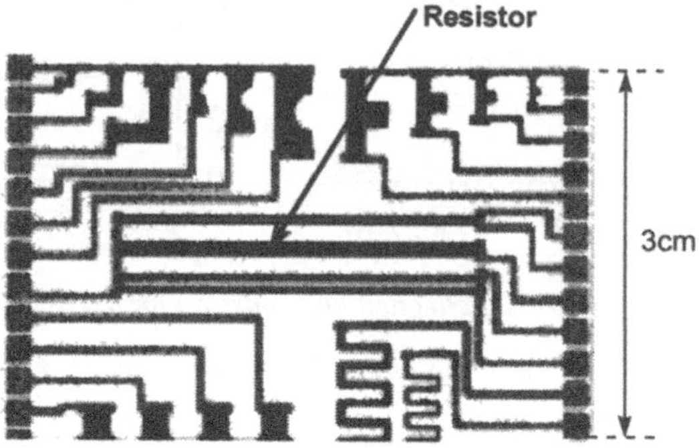

Figure 5. Hybrid resistor taken for the experiment

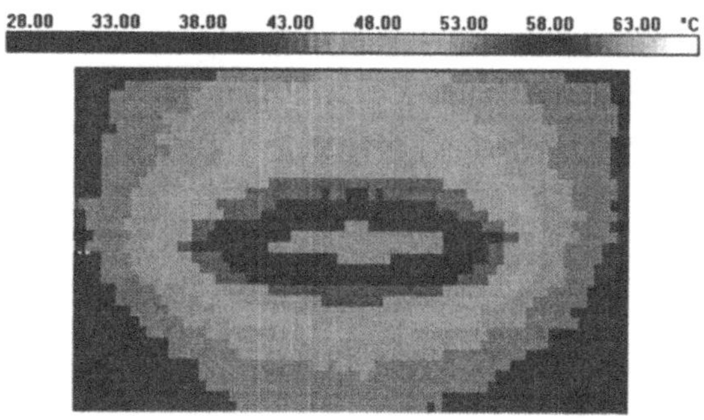

Figure 6. Hybrid resistor and temperature distribution while convective and radiated heat dissipated, P=1.7W, T_{max}=43°C

Using the results of the simulation we can approximate by exponentially both the power dissipated in the substrate and the total transfer coefficient versus temperature: $P_z=f(T_{max})$ and $a=f(T_{max})$ as:

$$P_z = P_0\left(\frac{\Delta T_{\max}}{\Delta T_0}\right)^m \qquad \alpha_{\max} = \alpha_0\left(\frac{\Delta T_{\max}}{\Delta T_0}\right)^m \tag{12}$$

The results for black body $(\varepsilon=1)$ using the *1/4 Law* for convection model are in Table 2. They confirm the high contribution of radiation in the total heat dissipation, especially for $T_{max} > 450K$. Above this temperature the heat is easier being removed by radiation than convection.

Table 2. $P_z=f(T_{max})$ and $\alpha_{max}=f(T_{max})$ approximation parameters

	Convection		Radiation		Convection +Radiation	
	P_0/α_0	m	P_0/α_0	m	P_0/α_0	m.
$P_z(T_{max})$	7.41	1.20	4.68	1.44	11.98	1.23
α_{max}	6.34	0.21	4.50	0.43	10.86	0.23

The next simulations were performed for grey bodies. Now the parameters of the non-linear model of convection are: α_0=5.5W/m^2 for ΔT_0=20°C[4-5]. Now the convective heat transfer coefficient varies according to equation (3).

Table 3. $P_z=f(T_{max})$ for gray bodies

	T_{max} [K]					
P_z[W/m]	ε=0	ε =0.2	ε =0.4	ε =0.6	ε =0.8	ε =1
50	482	434	412	397	387	379
100	640	532	492	467	451	438
150	790	611	557	525	504	488

The approximation $P_z \sim \Delta T^m{}_{Tmax}$ for e.g. $\varepsilon=0.4$ gives unexpected result - factor *m* is quite high indicating the high significance of radiation in heat removal.

$$P_z = P_0\left(\frac{\Delta T_{max}}{\Delta T_0}\right)^m \quad \text{where:} \begin{cases} P_0 = 5.11, m = 1.32\, for\, \varepsilon = 0.4 \\ P_0 = 8.72, m = 1.27\, for\, \varepsilon = 1.0 \end{cases} \tag{13}$$

For metals the maximum temperature versus power dissipated is presented in Table 4. We assumed the similar heat source in the middle of the substrate, but the different metals were used as substrates. We used an approximation of the emissivity by electrical resistivity, according to equation (10).

Table 4. $P_z=f(T_{max})$ for metals

	T_{max} [K]			
P_z[W/m.]	Al.	Fe	Fe-42Ni*	ε=0 (no radiation)
50	472	466	477	482
75	543	531	501	563
100	607	588	547	640

The curves below present the contribution of convection and radiation in heat removal process by using the model (4).

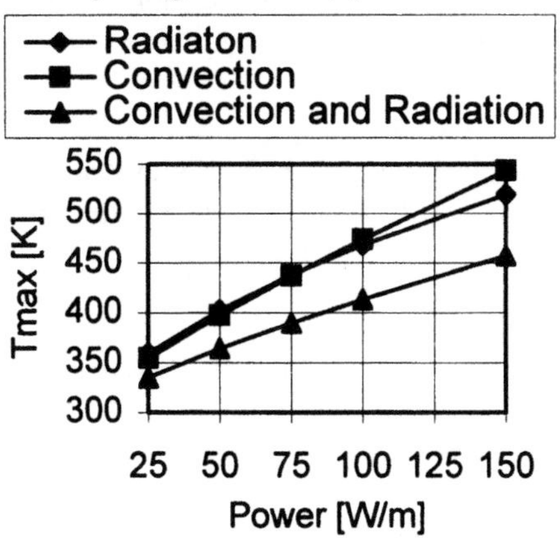

Figure 7. T_{max} versus power

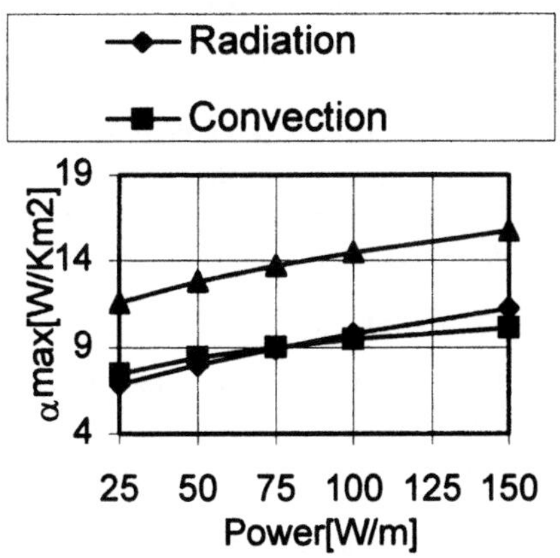

Figure 8. α_{max} versus power

For the model parameters described in the text the comparable amount of energy is dissipated both by convection and radiation, as shown in Figure 7. Heat transfer coefficient is non-linear (Figure 8). For temperature T_{max}>450K radiation begins to dominate over the convection in the heat removal process.

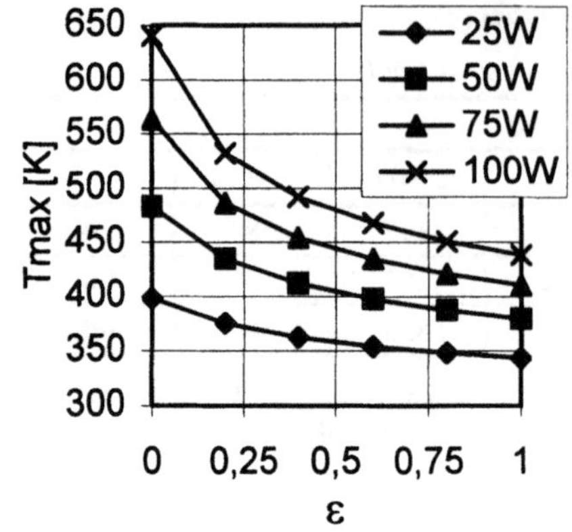

Figure 9. $T_{max}=f(\varepsilon)$ for gray bodies

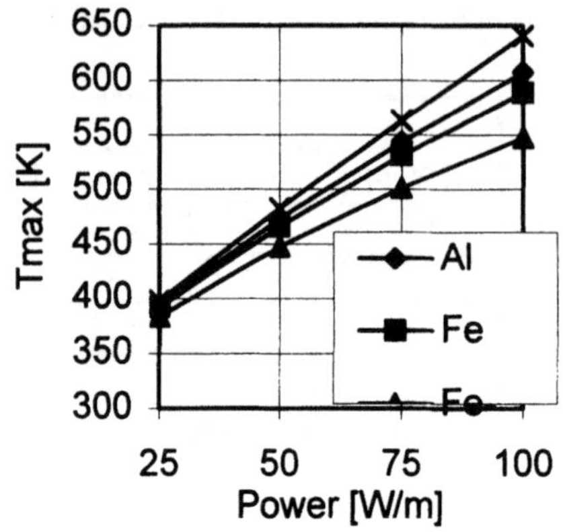

Figure 10. $T_{max}=f(P_Z)$ for metals compared to the convective heat dissipation only ($\varepsilon=0$)

As shown in Figure 9. the amount of energy dissipated to the ambient obviously decreases with lower emissivity. However the largest decay of T_{max} is in the emissivity range ε=0-0.4. This is a very practical result which indicates that there is no means to increase the emissivity to ε=1.

5. CONCLUSIONS

From tables and figures above we can come to some general conclusions:

- *Radiation and convection remove a comparable amount of heat in the temperature range accepted in microelectronics (e.g. below 200°C)*
- *The functions $P_z=f(T_{mean})$ and $a=f(T_{mean})$ are non-linear, and as far as the power dissipated by convection and radiation are comparable, radiation-convection and convection models give very similar variation of the T_{mean} and a over temperature (parameter m does not change very much). However the levels of energy dissipated are quite different*
- *From T_{mean}=150°C radiation dominates over convection*

The general conclusion drawn from the simulations for grey bodies is that, the decay of T_{max} is the largest *(70%)* for *ε=0-0.4,* which is the case in the practice *(Table 3, Figure 10).*

REFERENCES

[1] R. Siegel and J. Howell, *Thermal radiation heat transfer,*. New York, Hemisphere Publishing Corp, 1989.

[2] T. Burakowski, J. Giziński and A. Sala, *Promienniki podczerwieni* Warsaw, WNT, 1970 [in Polish].

[3] A. Sala, „*Radiacyjna wymiana ciepła*" Warsaw, WNT, 1982 [in Polish]

[4] B. Więcek and G. De Mey, *Evaluation of heat dissipation by convection for VLSI circuits,* Proc. TERMINIC'96 Conf., Budapest, Sept.25-27,1996

[5] B. Więcek,. *Quantitative approach into the heat transfer by convection in microelectronics with thermography measurements,* Proc. QIRT'96. Stuttgart, Sept. 2-5, 1996

12

MODELLING AND SYNTHESIS OF ELECTRO-THERMAL MICRODEVICES

Wojciech Wójciak, Andrzej Napieralski, Mariusz Orlikowski and Mariusz Zubert

Department of Microelectronics and Computer Science
Technical University of Łódź
al. Politechniki 11, 93-590 Łódź
POLAND

ABSTRACT

In this chapter the main steps of micro-transducers design flow will be discussed. First, the CMOS compatible MEMS technology is introduced. The example of micromachined chip is presented. Next, the electro-thermal model of chosen micro-device and the procedure of its synthesis are proposed. The results of the model verification are included. Finally, the application of the electro-thermal converter (ETC) to the power factor correction is described.

1. INTRODUCTION

There are plenty of sensors on the market but only the ones on which „smart" interfaces can be easily implemented will be mainly developed. Therefore the semiconductor sensors become very popular. Sophisticated computing circuit interfacing the sensor and controlling the actuator can be integrated in the same structure. Such a solution reduces significantly the cost of the whole system. Some of the microsensors, called the thermopile based, transform the energy of measured signal to heat and then to the electric voltage, exploring Seebeck phenomenon [1,2,3]. In IC, the most important problem is thermal separation of the sensor from the rest of the structure. For this reason the standard techniques for manufacturing the ICs with microsensors has been changed towards the new micromachined technologies.

2. CMOS COMPATIBLE MEMS TECHNOLOGY

The CMOS compatible, front side bulk micromachined technology was introduced in 1995 by CMP (Circuits Multi-Projets), Grenoble (France). Silicon wafers that have been processed through the standard CMOS process steps are processed further to release the microstructure areas formed from the thin films available in this process. The microsensor structure, which is typically formed by the polysilicon layer on the silicon dioxide, is released by etching the underlying silicon area. The thickness of the released microstructure is in the range of several μm and the length or the width is in the range of tens or hundreds μm. In such case the heat flux in the released structure can be sensed more efficiently. The microsensors are integrated together with the standard electronics on the same chip [4].

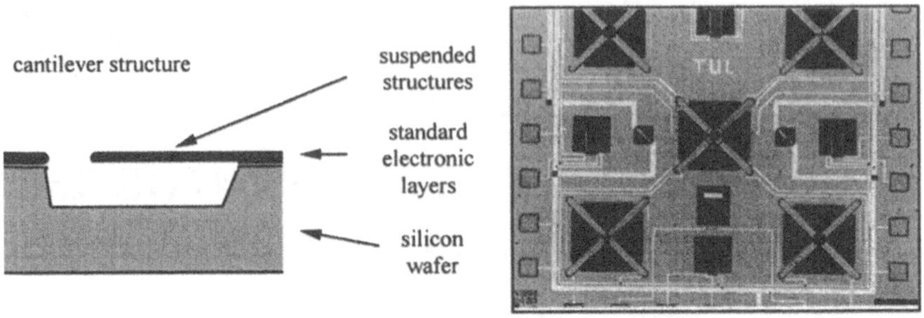

Figure 1. Simplified cross-section of suspended structure of electro-thermal converter and microscope view of micromachined chip

The VLSI ASIC chip (Figure 1) was designed using the MEMS CADENCE Design Kit for ES2 1.0µm CMOS technology. The chip contains infrared radiation sensors (IRS), electro-thermal converter (ETC), gas flow sensors (GFS), acceleration sensor (AS) and punctual light sources (PIX). The ETC device (Figure 2) consists of 13 serially connected aluminium-polysilicon thermocouples, and polysilicon heating resistor at the end of the cantilever. The device is thermally isolated from the bulk. The heat flowing from the heating resistor to the bulk is sensed by the thermopile. The efficiency of this device depends strongly on the proper and exact releasing the microstructure after etching post-processing step. If the structure is not totally released, the heating resistor should dissipate more heat in order to achieve the same value of the output voltage. Figure 3 shows the optical microscope view of fabricated electro-thermal converter. The heat dissipater is placed on the top of cantilever, thermopiles are placed along the cantilever.

3. ELECTRO-THERMAL MODELING AND SIMULATION OF ETC DEVICE

For designing such device, the behavioural simulation is required and it takes into account the following phenomena:

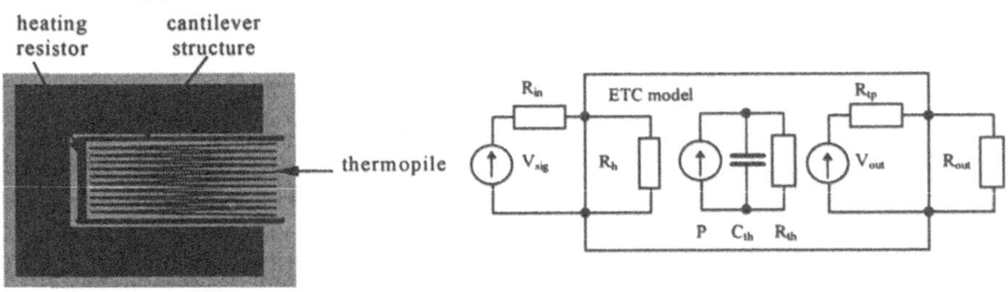

Figure 2. Simplified cross-section of suspended structure and model of electro-thermal converter

- heat generation in the heating resistor:
- heat transfer through the suspended structure which causes the temperature difference, for steady (1) and dynamic (2) state:

$$\Delta T = P \cdot R_{th} \tag{1}$$

$$c_p\rho\frac{\partial T}{\partial t} = \nabla(\lambda\nabla T) + q \tag{2}$$

ΔT - the temperature difference,
R_{th}- thermal resistance of the structure,
r - material density,
l - material thermal conductivity,
P - power dissipated in the heater
c_p -specific heat
T - temperature
q - generated heat density

- Seebeck phenomenon - thermopile voltage generation (without load)

$$V_{out} = N\cdot(\alpha_1 - \alpha_2)\cdot\Delta T = N\cdot\alpha\cdot\Delta T \tag{3}$$

V_{out} - Seebeck voltage,
N - number of thermocouples
α_1, α_2 - Seebeck coefficients of the thermocouple materials

- thermal changes of the heater and thermopile resistance
- heat flow to the ambient

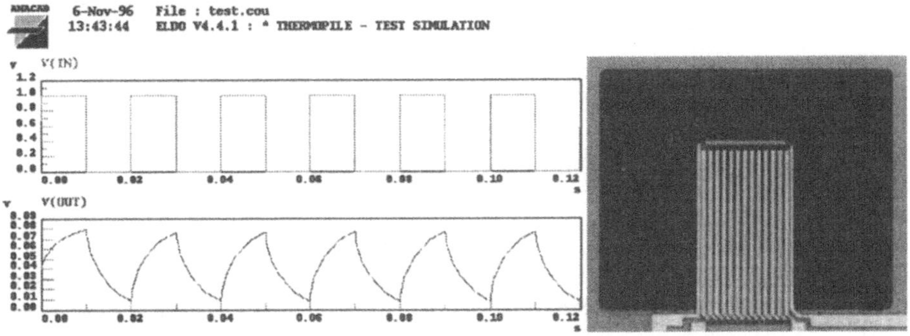

Figure 3. The thermopile simulation and microscope view of ETC device.

The HDL-A(TM) (Analogue extension of the Hardware Description Language) [5] model describes the behaviour of the ETC device. It includes also the differential equations of the heat transfer in the suspended structure. The structure of the device shown on Figure 3, allows the use of one dimensional equations. Simulated ETC structure was divided into a few elements and each of them was described by the difference equations, using FDM method. The electro-thermal simulations were performed with the ELDO-ANACAD(TM) simulator [6,7]. An example of thermopile response for the periodic pulse voltage force is illustrated on Figure 3. Other electro-thermal simulations show that such a kind of device cannot be treated as a lumped-parameter system (temperature distribution is very non-linear). Simulations with the HDL-A(TM) model and measurement results are shown on Figure 4.

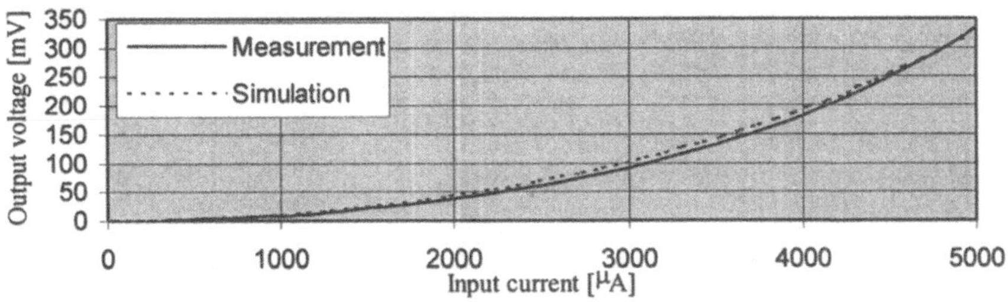

Figure 4. Measured ETC transfer characteristic $V_{out}=f(I_{in})$ compared with simulation results using HDL-A™ model.

4. SYNTHESIS OF ETC FROM HARDWARE DESCRIPTION

The ETC transducer can be automatically synthesized when its construction and technology in form of parametrized standard cells (P-Cell(TM)) is defined. The behavioural description should include the user-definable parameters like: sensitivity (S_{tp}), detectivity (D), thermal constant (τ_{th}), etc. and electrical and thermal properties of the particular device layers and geometric data restricted by design rules. Additionally, input and output resistance of the external circuit should be given for the purpose of determining the area of the heating resistor and the number of thermocouples (N). Finally, geometric parameters of the ETC are obtained and its layout generated.

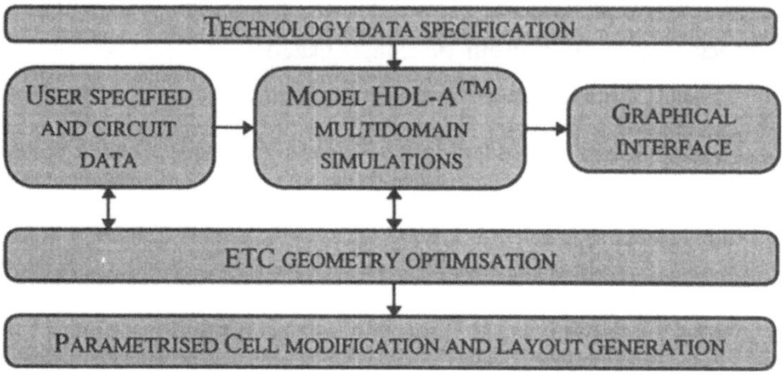

Figure 5. Synthesis procedure of the ETC transducer

For the ETC device with the thermopile width of W, consisting of N thermocouples of length L and width w, the set of equations (4), (5), (6) describes its main parameters.

$$S_{tp} = \frac{Vout}{P} = \frac{N \cdot \alpha}{G_{th}} \tag{4}$$

$$D = S_{tp} \cdot \sqrt{\frac{L \cdot W}{4 \cdot k \cdot T \cdot R_{tp}}} \tag{5}$$

$$\tau_{th} = \frac{C_{th}}{G_{th}} \tag{6}$$

Processing equations (4), (5), (6) with all technological and user-definable restrictions applied, the following set of equations for simplified ETC model can be obtained:

$$D = L \cdot c_{Det} \tag{7}$$

$$S_{tp} = \frac{L}{w} \cdot c_{Stp} \tag{8}$$

$$R_{tp(T=T_0)} = \frac{N \cdot L}{w} \cdot c_{Rtp} \tag{9}$$

where: c_{Det}, c_{Stp}, c_{Rtp} - constants,

The algorithm for ETC synthesis is presented on Figure 6.

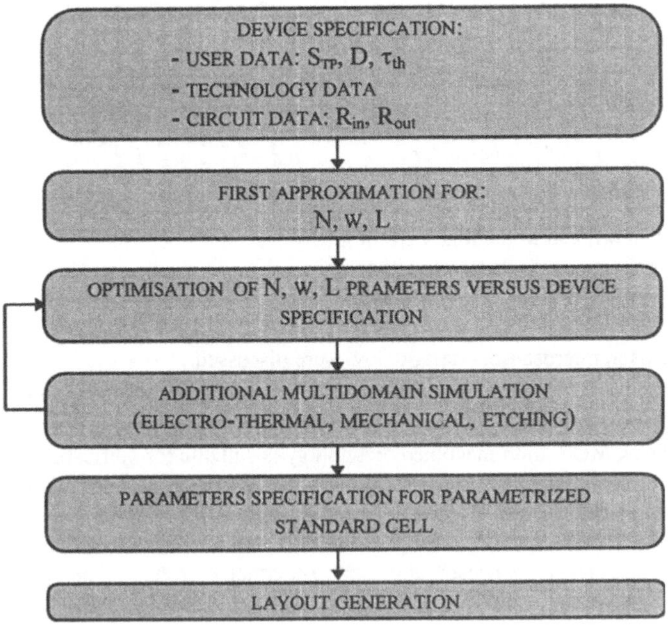

Figure 6. The ETC synthesis algorithm

5. POWER APPLICATION OF THE ETC DEVICE

The are two main application of the ETC device for power circuits. The first one is the heat flow meter for the purpose of thermal monitoring of semiconductor structure. The second one is the true RMS meter and active power transducer of distorted signals. Power factor correction systems uses mentioned transducers for the proper calculation of compensator control signals. Power factor correction (PFC) is successfully performed when the shape of the supply current is the same as the shape of the supply voltage and the RMS of the supply current depends only on the value of the active power consumed by the load [8,9]. In the case of highly distorted load current caused by PWM controlled power switch, reactive power can be compensated by the circuit acting as a controlled current source (Figure 7). The power meters with high accuracy are necessary for PFC process. Those are thermopile based. In MEMS technology the active power transducers, can be integrated with control chip on the same structure.

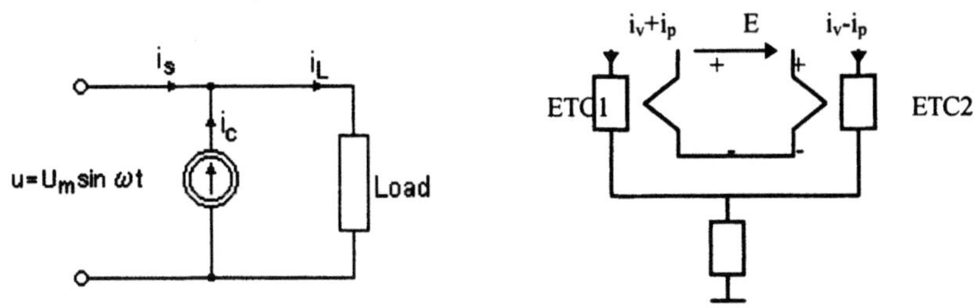

Figure 7. Simplified power factor correction circuit and active power transducer

The active power transducer uses an arrangement of two electro-thermal converters (ETC) in order to obtain output voltage proportional to the active power of input signals. Output voltage of the power transducer is described by equations (10):

$$E = c\frac{1}{T}\int_0^T (i_p + i_v)^2 - (i_p - i_v)^2 dt = c\frac{1}{T}\int_0^T 4 i_p i_v dt = c_1 \frac{1}{T}\int_0^T uidt = c_2 P \qquad (10)$$

where:
i_v - current proportional to the load current i_l,
i_p - current proportional to the load voltage u

6. CONCLUSIONS

The main steps in microsensors design-flow were discussed.

The possibility of designing the micro-electro-mechanical structures (MEMS) using standard CMOS technology with one additional mask-less step of etching has been presented. The CMOS micromachined technology is suitable for fabricating micro-electro-thermal converters, which can be integrated with the measuring and control circuitry on the same chip. The design of the VLSI ASIC chip fabricated in the CMOS MEMS technology was presented. The chip contains three kinds of microsensors: infrared radiation sensor, gas flow sensor and electro-thermal converter. The microsensors are based on the thermopile devices and explore the Seebeck phenomenon in the CMOS structure.

The modelling and simulation of the ETC device using HDL-A behavioural description was also introduced.

The procedure for the synthesis of ETC transducer from hardware description was proposed.

One of the possible applications is the power transducer for the power factor correction circuit.

REFERENCES

[1] S.M. Sze, *Semiconductor Sensors.* A Wiley-Intersc. Publ. 1994

[2] R. Lenggenhager, H. Baltes and T. Elbel, *Thermoelectric infrared sensors in CMOS technology*, Sensors and Actuators A, 37-38 (1993) pp. 216-220

[3] H.J. Verhoeven, *Smart Thermal Flow Sensors*, Ph.D. thesis, TU Delft, 1996

[4] First MPW MEMS run done at CMP, 10 projects. *CMP press release*, March'96

[5] J.M.Bergé and J.Rasilland, *Modeling in Analog Design*, Kluwer Academy Publishers 1996

[6] *HDL-A(TM) Language Reference Manual*, Revision 2.0, *ANACAD(TM) 1996*

[7] *HDL-A(TM) User's Manual, Revision 2.0, ANACAD(TM) 1996*

[8] N. Kuster, W. Moore, *On the definition of the reactive power under nonsinusoidal conditions*, IEEE trans. on Power. Appl. Syst. vol. PAS-99 nr3 1980.

[9] H. Supronowicz, *Poprawa współczynnika mocy w układach przekształtnikowych*, WNT Warszawa 1981 (in Polish)

PART III

Microsystems and Neural Networks

13

LAYOUT OPTIMIZATION OF CMOS PHOTOTRANSISTORS

[1]M. Moreno, [1]S.A. Bota, [1]R. Holgado, [1]A. Herms and [2]J. Calderer

[1]*EME-Departament de Física Aplicada i Electr nica*
Universitat de Barcelona. Av. Diagonal, 645-647. 08028 Barcelona
SPAIN

[2]*Departament d'Enginyeria Electr nica*
Universitat Polit cnica de Catalunya. C Gran Capitan s/n. 08034 Barcelona
SPAIN

ABSTRACT

A photodetector based on a vertical bipolar transistor, implemented using standard CMOS technology, was studied. The dc response of the photodetector was related to its layout. The contributions corresponding to area and perimeter were determined. Our results, allowed us to determine the minimum distance between two adjacent photodetectors so as to avoid optical cross-talk. Moreover, our findings can be used in improving the physical design of photodetectors and in device modelling.

1. INTRODUCTION

Given its low cost and potential for Very Large Scale Integration, silicon is an attractive material for the integration of photodetectors and signal-processing circuits [1]. Applications range from receiving elements used in optical communications systems [2-3] to vision chips, formed from a matrix of photodetectors, used in the transduction of images to electrical signals [4]. For this purpose, the concept of photoASIC has been developed, to denote VLSI integrated circuits involved in optical applications [4-5].

The aim of this work is to analyze the characteristics of one of the basic devices suitable for photoASIC implementation: the phototransistor. We seek to establish the photodetector device performances of various layout geometries in order to develop a simple method which can separete their area and perimeter contributions. An understanding of their respective contributions is important in photoASIC design in predicting the photocurrent value and also its influence on dynamic properties [6]. Integration of pixel arrays means not only definning array architecture but also evaluating the optical crosstalk effect occurring between two pixels. An analytical model to study optical crosstalk effects was also implemented and these results are discussed.

2. CMOS PHOTODETECTORS

When a standard digital CMOS process is employed, several structures, which are mainly based in junction devices, can be used as light sensitive elements. The device used in this study was the parasitic pnp bipolar junction transistor formed by a p^+ diffusion (emitter), an n-well (base) an the p substrate (collector) [6]. A schematic cross-section of the resulting phototransistor is shown in Figure 1. Note that this structure can also work as a photodiode assuming that the base is the n-port and the collector is the p-port.

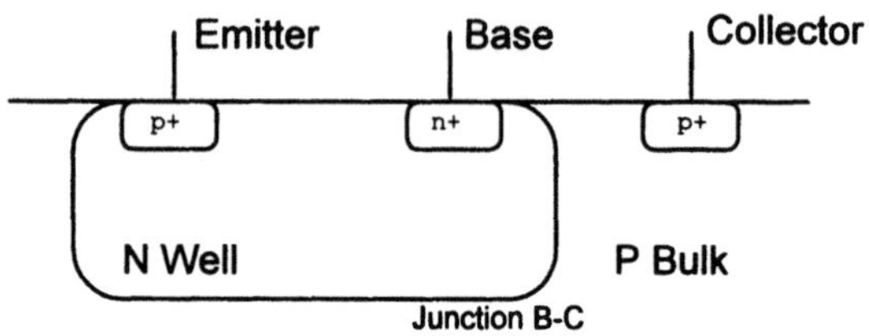

Figure 1. Photodetector cross-section

When the phototransistor is biased in the active forward region, the incident light generates electron-hole pairs in the reverse-biased base-collector junction that induce a current (base current); the area and perimeter of the base determine the maximum value of the photogenerated current. This base current is amplified by the gain factor ($1+\beta_F$), where β_F is the forward gain of the bipolar transistor.

3. DEVICE CHARACTERISTICS

The analyzed devices were implemented using a standard CMOS n-well process with one polysilicon and two metal layers. The depth of the well was 4.5 um. To analyze the influence of the phototransistor geometry, two series were fabricated. In the so-called R series, the area of the well (base) was kept constant, while the perimeter was changed. The C series consisted of square devices of differing area and perimeter. These layout parameters are presented in Table 1. The series RM and CM correspond to devices covered by a metal 2 layer, so as to surpress the carrier photogeneration under the well, thereby allowing the study of the side effects which are directly dependent on the perimeter. The area and perimeter of the emitter were chosen in order to ensure that the bipolar transistor was working in the linear region.

Table 1. Layout parameters (units are in microns)

R and RM Constant area = 8100 μm^2		**C and CM Square pixel**	
$B_1 = 90 \times 90$	$B_5 = 162 \times 50$	$C_1 = 175 \times 175$	$C_5 = 60 \times 60$
$B_2 = 101 \times 80$	$B_6 = 202 \times 40$	$C_2 = 125 \times 125$	$C_6 = 50 \times 50$
$B_3 = 115 \times 70$	$B_7 = 270 \times 30$	$C_3 = 80 \times 80$	$C_7 = 40 \times 40$
$B_4 = 135 \times 60$	$B_8 = 324 \times 25$	$C_4 = 70 \times 70$	$C_8 = 30 \times 30$

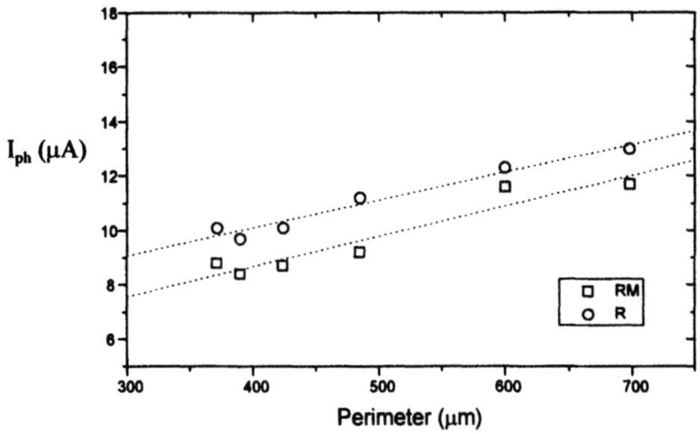

Figure 2. Perimeter dependence of I_{ph} corresponding to R and RM devices

4. EXPERIMENTAL RESULTS

The resulting parasitic transistor, presented in Figure 1, was characterized. The reverse-biased well-substrate junction under illumination collected a photocurrent, which was amplified by the transistor, the measured gain was constant across more than five decades of incident light intensity, $\beta_F = 45$. The spectral response curve showed a peak centered at 750 nm [7].

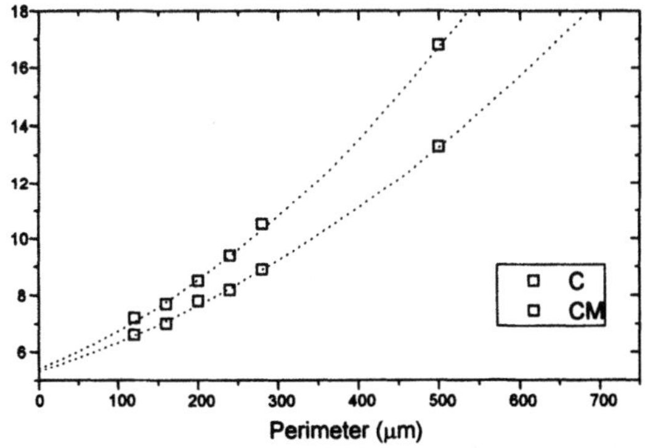

Figure 3. Perimeter dependence of Iph corresponding to C and CM devices

Table 2

Device	Curve Fitting (Iph in μA, P in μm)
R	Iph = 6.02 + 1.1 10-2 P
RM	Iph = 4.23 + 1.0 10-2 P
C	Iph = 5.41 + 1.1 10-2 .P + 3 10-5 P2
CM	Iph = 5.32 + 9 10-3 P+ 1.4 10-5 P2

Measurements were performed under light conditions corresponding to solar light of 50 W/cm^2. It was observed that the absolute current values after amplification, I_{ph}, were transistor size dependent. In Figure 2, the linear dependence of I_{ph} versus the base perimeter corresponding to the R and RM devices is shown. Figure 3 shows the dependence of I_{ph} on the base-perimeter, P, for the C and CM samples. In Table 2, the values of the curve fitting expressions are presented.

5. DISCUSSION

It is important to note that series CM and RM showed responses to light, assuming that the metal2 layer used in these series, prevents the passage of light through the n-well. It is inferred that the response of a pixel is not only dependent on the n-well area. In the framework of VLSI technologies, pixel area and spacing are comparable in size with the diffusion length of minority carriers, L_n. So, it is expected that a significant contribution of the total collected current is generated by the incident light in the neighborhood of the well.

From the analysis of the results presented in Table 2, we can consider that the total photocurrent is the sum of the contributions coming from the three regions (Figure 4).

$$I_{ph} = I_I + I_{II} + I_{III} \tag{1}$$

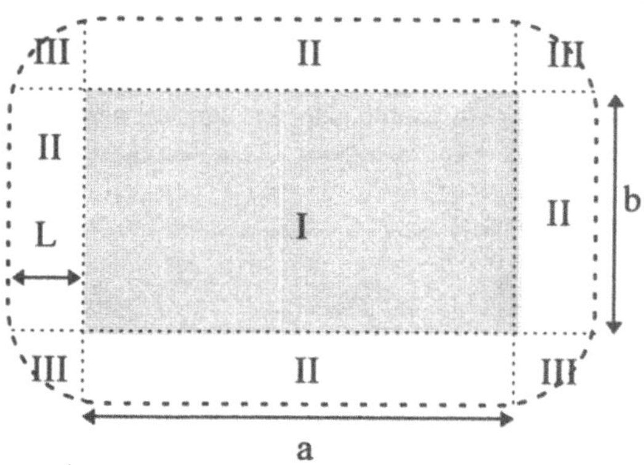

Figure 4. Three regions model. a·b and 2(a+b) correspond to phototransistor area and perimeter

Region I corresponds to the n-well area, A_I, its contribution to the current, and I_I, is related to carriers photogenerated beneath the surface of the well. The current I_{II}, generated in Region II, is proportional to the well perimeter. The photocurrent, I_{III}, generated in the corners (Region III) is independent of the well size and can be related to the independent term in Table 2.

The difference between $I_{ph,R}$ and $I_{ph,RM}$, corresponding to equivalent samples of series R and RM is constant (≅1.3 μA), and can be related to the fraction of the photocurrent generated by the light incident on the n-well area. We assume that in samples RM, the response comes from the light incident on areas II and III, A_{II} and A_{III}:

$$A_{II} + A_{III} = 2(a+b)L + 4L^2 \tag{2}$$

L is defined in Figure 4. This area depends only on the well perimeter. From the current measurements presented in Figure 2 and taking into account that:

$$\frac{I_{ph,R} - I_{ph,RM}}{A_I} = \frac{I_{ph,RM}}{A_{II} + A_{III}} \tag{3}$$

In knowing A_I and from eq. (2), one can obtain the parameter L, here L=65 μm. Optical crosstalk effects will appear if the spacing between two pixels is inferior to 2L.

The quadratic dependence of I_{ph} vs P, found in Figure 3 corresponding to C devices, is due to the fact that both area and perimeter, change from sample to sample. The quadratic behaviour for CM devices suggests that a fraction of the incident light (≅60%) reaches the collector through the metal 2 layer and the n-well, this means that the metal layer is not completely opaque. Taking this into account, the parameter L is revised, and a corrected value of L≅45 μm is obtained.

6. DEVICE MODELLING

An analytical model was developped to study the influence of the technological parameters on the photogenerated current and on optical crosstalk. The photogenerated current was calculated by solving the diffusion equation of minorities on a 2D geometry taking into account bulk and surface photogeneration and recombination effects [8].

Figure 5a shows the simulated structure, where 2A and B are, respectevily, the witdh and thickness of the junction photodiode. Figure 5b represents the substrate photocurrent as a function of W, the illumination window, assuming different values of L, the minority carrier diffusion length.

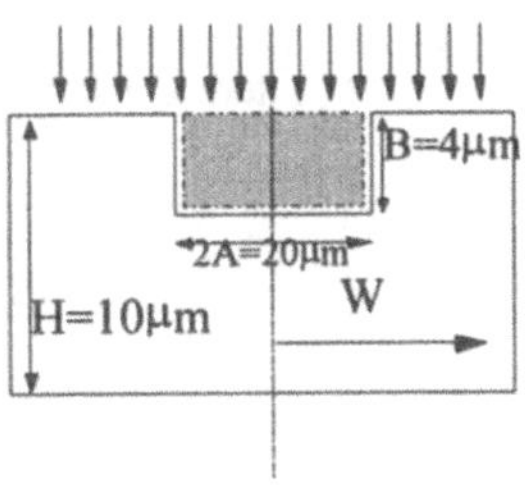

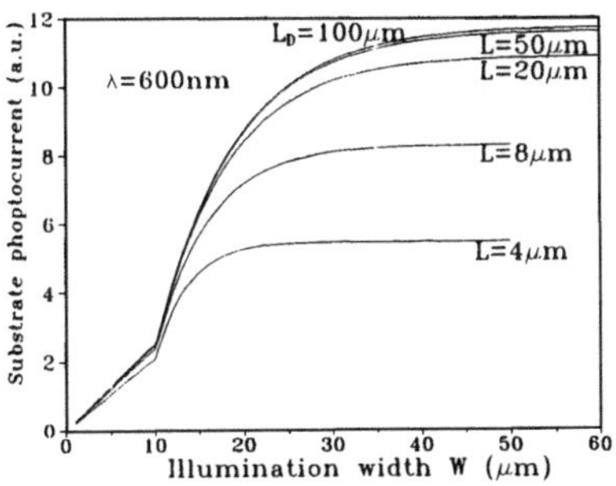

Figure 5. a) Simulated 2D structure *b)Effective illumination width*.

It was found that photocurrent saturation occurs when W reached a L-dependent threshold value. It is important to discusse the behaviour of plots corresponding to L=50 and 100μm, as a consequence of the geometrical parameters related to CMOS technology, the increase in L over 50 μm, had no influence in the value of the saturation current. In agreement with the experimental results, the simulated results suggest that an effective difussion length of L=40-50 μm must be expected in our CMOS structures.

Another immediate consequence of the simulated results was that some carriers can reach a neighbour photodiode and contribute to optical crosstalk effects.

7. CONCLUSION

A method for separating well-area and perimeter contributions has been determined. In which these two contributions to Iph could be measured for our target CMOS technology. The determination of these values is important because they might be related to a region close to the photodetector and which contributes to the total response. It will therefore be a major determinant of the minimum distance between two adjacent photodetectors, for instance in forming a pixel matrix, in order to avoid optical cross-talk or blooming effects.

ACKNOWLEDGEMENTS

This work has received the financial support of CICyT projects MIC89-035 and TIC96-1045.

REFERENCES

[1] *Special Issue on Solid-State Image Sensors*, IEEE Electron Devices, Vol. 38, n 5, 1991

[2] M. Yamamoto, M. Kubo and K. Nakao, *Si-OEIC with a Built-in-Photodiode*, IEEE Electron Devices, Vol. 42, n 1, pp 58-63, 1995.

[3] Y. Huang, *Optimized Integrated CMOS Optical Receiver for Optical Interconnects*, IEEE Proceedings-J, Vol. 140, n 2, pp 107-114, 1993

[4] J. Kramer, P. Seitz and H. Baltes, *Industrial CMOS Technology for the Integration of Optical Metrology Systems (photo-ASICs*", Sensors and Actuators A, n 34, pp 21-30, 1992

[5] H. Ando, S. Ohba, M. Nakai, N. Ozawa, K. Ikeda, T. Masuhara, T. Imaide, I. Takemoto, T. Suzuki and T.Fujita, *Design considerations and performance of a newMOS imaging device*, IEEE Electron Devices, Vol. 32, n 8, pp 1484-1489, 1985

[6] M. Moreno, *Fotodetectores realizados en tecnología CMOS y su aplicación a sensores ópticos integrados*, Tesis Doctoral, UPC, 1995 (in Spanish)

[7] J. Calderer, M. Moreno and M. Braam, *Integration of Phototransistors in CMOS circuits*, Sensors and Materials, Vol. 8, n 4, pp 199-208, 1996

[8] M. Moreno, J. Calderer and S. Bota, *2D analysis of a pn junction photoresponse and its application to CMOS photodetector arrays*, To be published in Solid State Electronics.

14

MULTILAYER PIEZOELECTRIC SENSORS ON THE BASIS OF THE PZT TYPE CERAMICS

[1]Dionizy Czekaj, [1]Julian Dudek, [1]Zygmunt Surowiak, [2]Aleksandr V. Gorish, [2]Yuri N. Koptev, [3]Aleksandr A. Kuprienko, [3]Anatoli E. Panich

[1]University of Silesia, Institute of Engineering Problems
2, Śnieżna St., Sosnowiec, PL 41-200,
POLAND

[2]Research Institute of Physical Measurements of the Space Agency
Moscow
RUSSIA

[3]Rostov State University, Scientific-Design-Technological Department "PIEZOPRIBOR"
5, Zorge St., Rostov on Don, SU 344104
RUSSIA

ABSTRACT

A model of multilayer piezoelectric sensor consisting of piezoactive layers alternating with thin metal electrodes is briefly reported. On the basis of the model a construction of the piezoelectric transducer was developed. Influence of temperature on the piezoelectric signal of such a sensor has been considered. As result of calculations a method of controlling the value of pyroelectric signal by right selection of sort, thickness and number of the ceramic and metallic layers constituting the piezoelectric element has been shown.

1. INTRODUCTION

It is a common knowledge that the piezoelectric ceramic sensors as well as the thin film sensors [1-7] have gained widespread application in automatic control systems used in modern technology (e.g. aircraft industry, automotive industry, machine-building industry, space engineering etc.). The physical parameters measured by piezoelectric sensors include a wide range of mechanical and electrical quantities [1,2]. Among others, they are used for measuring quick-varying pressures, strains, accelerations, vibrations, as well as characteristics of impact, force, flow rate and other physical quantities. The piezoelectric ceramic sensors are typified by exceptionally good operating characteristics (almost ideal for same purposes), they are proof against the action of multicycle load, they have simple construction, they are reliable and cheap, they are of small dimensions and they do not need external power sources.

Piezoelectric ceramic transducers can operate under tough operating conditions (e.g. within the wide range of temperatures, vibrations, acoustic noises, linear accelerations, mechanical impacts and water hammers, at high temperature gradients as well as in corrosive and cryogenic media). In many cases they often undergo action of few physical disturbing factors at the same time. It is considered as a rule, that the piezoelectric ceramic sensor ought to be characterised by the mechanical strength greater than the strength of the whole construction or the unit where the sensor is mounted on. To determine the ultimate load of the sensor the last ought to „live" longer than construction of the device (machine, assembly etc.). However, the piezoelectric ceramic sensor ought to be typified by high stability of the properties during the whole period of operation.

Taking into account a wide spectrum of possible external actions on the piezoelectric ceramic sensor as well as possible response of the piezoelectric element to a particular type of the actions, the main problem, during the engineering development of the sensor, is separation of the legitimate signal and decreasing of the measurement error.

It is shown in the present paper that, taking into account the theoretical model of the multilayer piezoelectric sensor [8], one can build the sensing element of the piezoelectric transducer with negligible influence of the pyroelectric effect at any temperature changes.

2. THEORETICAL MODEL

The multilayer piezoelectric ceramic sensor in the form of disk-shaped layers of polarised ceramics separated by metal layers having thickness of l_1 and l_2, respectively (Figure 1), was taken under consideration.

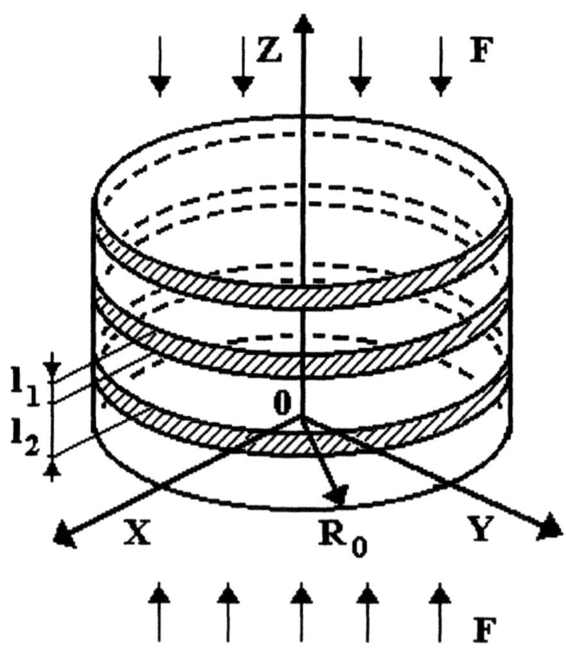

Figure 1. Multilayer piezoelectric transducer; l_1 - thickness of the ceramic layer; l_2 - thickness of the metallic layer; R_0 - radius of the piezoelectric ceramic sensor in the layer interface plane

The term of the relative thickness or, in other words, the volume density (m_i) was introduced as follows:

$$m_i = \frac{l_i}{2l} \tag{1}$$

where:

2l - the layers system period ($2l = l_1+l_2$).

It is worth noting that the basic and the only assumption that has been chosen [8] is as follows: the layers system period 2l has been taken as considerably smaller than the cross-sectional dimensions of the piezoelectric ceramic sensors in the layer interface plane. In other words, the one-dimensional problem was under investigation.

Such a piezoelectric system consisting of parallel layers one can consider as a polar texture with ∞·m symmetry and describe by means of the characteristic equations for the thermo-electroelastic anisotropic system [9]:

$$D_i = \varepsilon_{ij}^{\sigma,\theta} E_j + d_{i\alpha}^{\theta}\sigma_\alpha + p_i^{\sigma}\theta \tag{2}$$

$$\xi_\beta = d_{k\beta}^{\theta} E_k + S_{\alpha\beta}^{E,\theta}\cdot\sigma_\alpha + \alpha_\beta^{E}\theta \tag{3}$$

$$\Sigma = p_i^{\sigma} E_i + \alpha_\beta^{E}\sigma_\beta + \frac{\rho C_p^E}{T}\theta, \tag{4}$$

where:

$i, j = 1, 2, 3$;

$\alpha, \beta = 1, 2....6$;

D_i - electric induction vector components;

E_j - electric field intensity components;

σ_α - mechanical stress tensor (matrix notation: α=1,2....6);

θ - temperature change; ($\theta = T - T_0$, where T - current temperature of the piezoelectric element, T_o - initial temperature);

$\varepsilon_{ij}^{\sigma,\theta}$ - dielectric permittivity tensor components measured under constant stress (σ = const, $\theta = 0$);

$d_{i\alpha}^{\theta}$ -piezoelectric modulus tensor components measured under isothermal conditions ($\theta = 0$);

ξ_β - elastic deformation tensor components (matrix notation: β = 1,2...6);

Σ - change of entropy; ($\Sigma = S - S_o$, where S_o - entropy of the initial state; S - current entropy of the system);

$S_{\alpha\beta}^{E,\theta}$ - elastic mechanical compliance (elastic constant) tensor components measured under constant electric field (E = const, $\theta = 0$);

p_i^{σ} pyroelectric coefficient of the mechanically free crystal describing the total pyroelectric effect measured in the absence of stresses ($\sigma = 0$);

α_β^E - tensor components of coefficient of thermal expansion measured under constant electric field (E = const);

C_p^E - specific heat at constant pressure (p = const) and constant electric field (E = const);

ρ - density of the piezoelectric material.

To describe behaviour of such heterogeneous layered piezoelectric system under the simultaneous action of the electric, mechanical and thermal fields the averaged physical constants, which depend on properties of the layers and their thickness were used.

The method of solving the system of equations (2)-(4) was based on the assumption that one can find co-ordinated dependence of the 10 thermodynamical variables for the one-dimensional model of the sensor. Then the effective properties of the piezoelectric sensing element i.e. the ones appearing at the macroscopic experiment can be obtained by integration over a thickness of the multilayer piezoelectric sensor taking into account averaged microfields [8].

As result of calculations one can obtain the following set of the effective tensors of physical properties of the piezoelectric ceramic sensing element [8]:

$$(\varepsilon_{33}^{\sigma,\theta})^* = \frac{\varepsilon_{33}^{\sigma,\theta}}{m_1}\left[1 - \frac{2m_2(k_{31})^2}{m_1(1-\nu)/(ES_{11}^{E,\theta}) + m_2(1-\nu_{12}^E)}\right]; \tag{5}$$

$$(\varepsilon_{11})^* = \infty; \tag{6}$$

$$(d_{33}^\theta)^* = d_{33}^\theta - \frac{2d_{31}^\theta(\nu + S_{13}^{E,\theta}E)}{(m_1/m_2)(1-\nu) + (1-\nu_{12}^E)S_{11}^{E,\theta}E;} \tag{7}$$

$$(d_{31}^\theta)^* = d_{31}/\left[m_1 + m_2(1-\nu_{12}^E)S_{11}^{E,\theta}E/(1-\nu)\right]; \tag{8}$$

$$(d_{15})^* = m_1 d_{15}; \tag{9}$$

$$(S_{11}^{E\theta})^* = \frac{S_{11}^E\left[m_1 + m_2E(S_{11}^2 - S_{12}^2)/S_{11}^E(1-\nu^2)\right]}{\left[m_1 + m_2(S_{11}^E - S_{12}^E)E/(1+\nu)\right]\cdot\left[m_1 + m_2(S_{11}^E + S_{12}^E)/(1-\nu)\right]}; \tag{10}$$

$$(S_{12}^{E\theta})^* = \frac{S_{12}^E\left[m_1 - m_2\nu E(S_{11}^2 - S_{12}^2)/S_{12}^E(1-\nu^2)\right]}{\left[m_1 + m_2(S_{11}^E - S_{12}^E)E/(1+\nu)\right]\cdot\left[m_1 + m_2(S_{11} + S_{12})E/(1-\nu)\right]}; \tag{11}$$

$$(S_{33}^{E\theta})^* = m_1S_{33}^E + m_2/E + \frac{2m_1m_2(\nu/E + S_{13}^E)\left[2k_{31}^2(\nu/E + S_{13}^E)/(1-\nu_{12}^E) - (\nu/E - S_{13}^E) - 2d_{31}d_{33}\varepsilon_{33}^\sigma\right]}{(1-\nu)\left[m_1 + m_2(1-\nu_{12})S_{11}^E E/(1-\nu)\right]\cdot\left[1 - 2k_{31}^2/(1-\nu_{12}^E)\right]}; \tag{12}$$

$$(S_{44}^{E\theta})^* = m_1S_{44}^E + m_2/G; \tag{13}$$

$$(S_{66}^{E\theta})^* = (S_{66}/G)/(m_1/G + m_2S_{66}); \tag{14}$$

$$(p_3^\sigma)^* = p_3^\sigma - \frac{2m_2d_{31}^\theta(\alpha_1 - \alpha)E/(1-\nu)}{m_1 + m_2(S_{11}^{E,\theta} + S_{12}^{E,\theta})E/(1-\nu)}. \tag{15}$$

where:

$d_{33}^\theta, d_{31}^\theta$ - piezoelectric moduli;

$\varepsilon_{33}^{\sigma\theta}$ - dielectric permittivity;

k_{31} - electromechanical coupling coefficient,

p_3^σ - pyroelectric coefficient;

α_1 - coefficient of thermal expansion of ceramics;

α - coefficient of thermal expansion of metal;

$S_{11}^{E\theta}, S_{12}^{E\theta}, S_{13}^{E\theta}$ - elastic constants of ceramics (elastic mechanical compliance);

$\nu_{12}^E = -S_{12}^{E,\theta}/S_{11}^{E,\theta}$;

E - Young's modulus,

G - shear modulus of the metal layer ;

ν - Poisson's ratio;

symbol (*) denotes effective values of the physical quantities.

One can see from equations 5-15 that effective physical parameters of the piezoelectric element depend on the material constants of the layers as well as their relative thickness.

3. RESULTS OF THE EXPERIMENT

Computer analysis of the pyroelectric coefficient of the piezoelectric sensing elements made of CTS-19 - type and PZT-5 - type ceramics as well as bismuth titanate one has shown that the crosswise effects influence the behaviour of the sensor. They contribute their share to an electric signal coming from strains caused by temperature. The sensors built on the basis of ceramics with small anisotropy of piezoelectric moduli and aluminium, 36HXTQ-type steel, titanium, brass or lead are characterised by decreasing the effective pyroelectric coefficient (p_3^σ)* with increasing in the contribution in question. Moreover, for the definite ceramic-to-metallic layer thickness ratio the pyroelectric coefficient gets value of zero.

One can see in Figure 2 the dependence of the pyroelectric coefficient of the sensor made from a piezoelectric ceramics - metal system on the relative thickness of the CTS-19 - type ceramics. The following metallic layers were used: 1 - aluminium, 2 - 36HXTQ-type steel and brass, 3 - lead, 4 - platinum, 5 - 32HKD-type steel.

One should choose the relative thickness of ceramics m_i taking into account its chemical composition. For the CTS-19 and PZT-5 based piezoelectric elements the relative thickness works out at 72 and 78% for aluminium, 48 and 63% for 36HXTQ-type steel 18 and 60% for brass, 42 and 52% for silver, 50 and 58% for tin, 30 and 36% for lead, respectively (percent is given in relation to the whole thickness of the ceramics-metal system). The thermal expansion coefficients for these systems fulfil the requirements: α_1 (ceramics)<α (metal). If platinum or 32HKD-type steel is used as a metal layer ($\alpha_1 \geq \alpha$ for these materials), the pyroelectric coefficient is always above zero and it increases with decreasing in the relative thickness m_i.

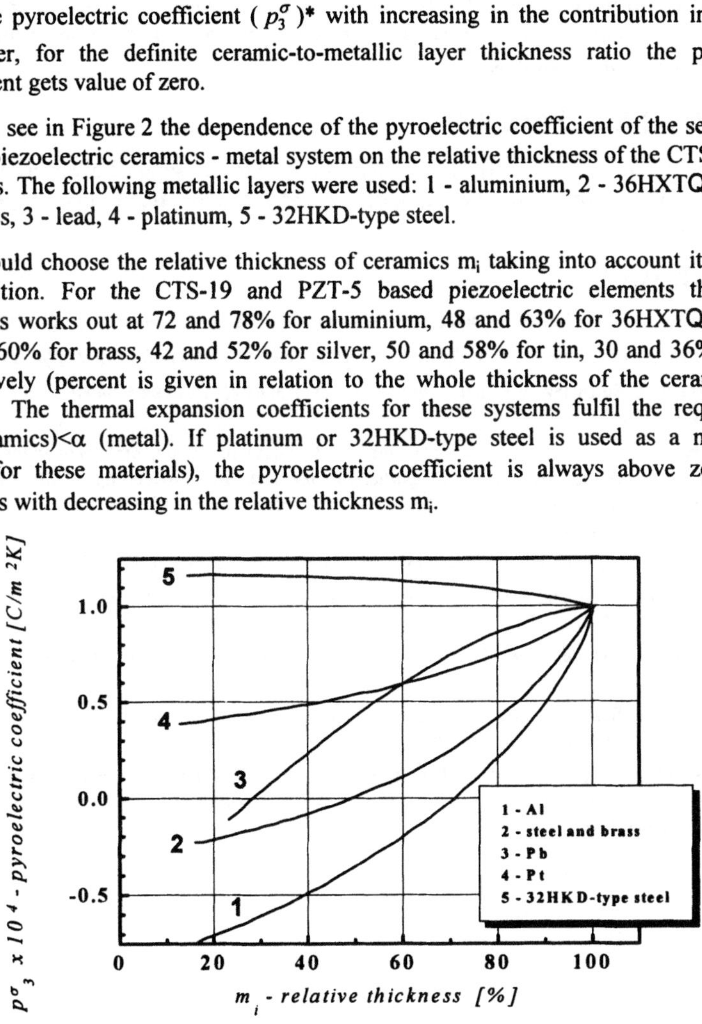

Figure 2. Dependence of the pyroelectric coefficient of the multilayer piezoelectric transducer on the relative thickness of the PZT-type layers

For the bismuth titanate ceramics the described above effect is not observed due to high anisotropy of the piezoelectric moduli - $\left|d_{33}^{\theta} / d_{31}^{\theta}\right| = |18|$. One can conclude that the crosswise piezoelectric effects have small influence on pyroelectric coefficient.

It should be pointed out that the results mentioned above were obtained for the rigid connection between the ceramic and metallic layers.

4. CONCLUSIONS

On the basis of the mathematical model describing all important characteristics of the piezoelectric ceramic element the construction of the multilayer piezoelectric ceramic sensor was developed. The results of investigation on construction of the piezoelectric sensors make it possible to conclude that one can eliminate the temperature influence on the piezoelectric signal by means of right selection of ceramics and metal as well as their thickness and number of layers constituting the piezoelectric element. The proposed method of controlling value of the pyroelectric signal makes it possible to decrease the thermal noise. Therefore, one can consider it as a significant contribution to high stabile piezoelectric sensor construction.

ACKNOWLEDGEMENTS

One of the authors (D. Czekaj) deeply appreciates financial support given by the Polish State Committee for Scientific Research (KBN) within framework of the grant N° 8 T11B 079 10.

REFERENCES

[1] Z. Surowiak, D. Czekaj, V.P. Dudkevich. A.A and Bakirov:,*Elektronika*, 1, 12 (1994)

[2] Z. Surowiak, D. Czekaj, A.A. Bakirov and V.P. Dudkevich, *Thin Solid Films*, 256, 226, (1995)

[3] D. Czekaj, Z. Surowiak, V.P. Dudkevich and A.A. Bakirov, *Akustyka Molekularna i Kwantowa*, 15, 43, (1994)

[4] D. Czekaj, Z. Surowiak, A.A. Bakirov and V.P. Dudkevich, *Piezoelectric properties of the thin PZT-type films obtained by r.f. sputtering method*, In: Proceedings "8th Piezoelectric Conference PIEZO'94", 5-7 October 1994 Zakopane. Tele and Radio Research Institute, Warszawa 1995, p.289.

[5] Z. Surowiak, J. Dudek and M. Łoposzko, *Inżynieria Materiałowa*, 6, 123, (1992)

[6] Z. Surowiak, D. Czekaj, A.M. Margolin, E.V. Sviridov, V.A. Aleshin and V.P. Dudkevich, *Thin Solid Films*, 214, 78, (1992).

[7] Z. Surowiak, J. Dudek, Yu.I. Goltzov, I.A. Bugayan and V.E. Yurkevich, *J.Mater.Sci*, 26, 4407, (1991).

[8] D. Czekaj, Z. Surowiak, A.V. Gorish, A.A. Kuprienko and A.E. Panich, Kwartalnik Elektroniki i Telekomunikacji, 42 , 2, 227 (1996)

[9] D. Berlinkur, D. Kerran and G. Zhaffe, *Piezoelektricheskie i piezomagnitnye materialy i ikh primenenie v preobrazovatelakh,* In: Fizicheskaya akustika, t.1 (A). Ed. U. Mezon. Mir, Moskva 1966, 204.

15

ARTIFICIAL NEURAL NETWORK MIXED-SIGNAL PROTOTYPE SYSTEM FOR MODEL PARAMETER IDENTIFICATION

Andrzej Materka, Pawel Pełczynski and Michał Strzelecki

Institute of Electronics
Technical University of Lodz
Stefanowskiego 18, 90-537 Lodz
POLAND

ABSTRACT

Principles of model parameters identification by means of ANN-like approximators of multivariable mappings are reviewed briefly. The technique offers very high speed and lower noise-induced error compared to traditional methods. However, the ANN training becomes time consuming and somewhat uncontrollable at large range of unknown model parameters and high accuracy level required. It is proposed to use a modular, classifier-approximator architecture to get rid of this deficiency. Results of computer simulation are presented to illustrate the idea. Neural-network-based hardware system design for model parameter identification is proposed.

1. INTRODUCTION

Parametric modelling of dynamic systems is a widely used technique of describing physical systems of interest. Searching for fast and reliable methods of model parameter estimation based on system observation is a crucial problem in many applications [1]. A model of a given system can be a set of differential/difference/algebraic equations of constant coefficients (parameters). Assume model structure is known but actual values of the parameters are unknown. To identify the parameters, the system is excited by a predetermined test signal (stimulus), specified either in time or frequency domain depending on the nature and properties of the system. The system response $y=(y_1,\dots,y_k)'$ to the stimulus is measured, where (.)' denotes the vector transpose. It forms, either directly or after suitable preprocessing, the system observation vector $\varphi=(\varphi_1,\dots,\varphi_n)'$, $k\geq n$ and $\varphi\in\Phi\subset\Re^n$ The parameter identification problem is defined as follows: given the observation vector φ find estimates $\hat{\theta}=(\hat{\theta}_1,\dots,\hat{\theta}_p)'$ of the model parameters $\theta=(\theta_1,\dots,\theta_p)'$, $n\geq p$. The model parameters $\theta\in\Theta\subset\Re^p$ are assumed constant during the observation interval. For a fixed stimulus, the observation vector is a multivariable function of the model parameters $\{\varphi=f(\theta)\colon \Re^p\to\Re^n\}$ [1].

Some additional measures have to be undertaken to make the problem solvable. Namely, the stimulus signal definition, sampling moments selection (or input frequencies selection in the case of sine-wave testing) and the choice of the preprocessing algorithm have to be decided in a way which ensures that a unique inverse mapping $\{\theta = f^{-1}(\varphi): \Re^n \rightarrow \Re^p\}$ from the observation domain Φ to the parameter domain Θ exists [2]. It is assumed in this paper that the inverse mapping of interest exists and is unique.

2. PARAMETER ESTIMATION BY MEANS OF ANN APPROXIMATORS

It can be noticed that model parameter identification is the process of evaluating the vector value of the inverse mapping $f^{-1}(\varphi)$ given the observation vector φ. On the other hand, it is well known [3] that artificial neural networks (ANNs) are able to approximate any continuous compact-support function to any degree of accuracy. It was then postulated in [4] that ANNs can be used to approximate the mapping of interest $\hat{\theta} = g(\varphi) \cong f^{-1}(\varphi)$, provided they have been trained properly. The training itself is an iterative process which can take a substantial amount of computer time. However, it is done once only for a given model and its parameter range. After the training, in the recall mode, the ANN produces its response, i.e. model parameter values, in a very short time - just in one shot in the case of feedforward neural networks. High speed is therefore one of the advantages of the ANN-based parameter identification technique, compared to traditional techniques which typically employ iterative time-consuming computations [1], [5]. This feature makes the ANN-based method suitable for real-time applications.

Training the ANN approximator for model parameter identification involves minimising the differences between the actual and target <u>parameters</u>, whereas the model-tuning technique produces parameter values by minimising differences between the model and system <u>observations</u>. Therefore, the ANN trained on noisy observations becomes less sensitive to those input data which would normally affect much the LSE-fitted model parameters. This higher noise immunity when compared to traditional methods is another advantage of the ANN-based technique [6].

The novel technique discussed can be applied to any model: linear, non-linear, continuous- or discrete-time, provided that the mapping $\theta = f^{-1}(\varphi)$ exists and there is an ANN network able to approximate it with a tolerable error. The numerous applications range from electronic circuit parametric testing [7], through control engineering [8] to biomedical engineering [6]. The technique can also be used for indirect measurements. Part of the current research work, aimed at further development of the technique, relates to investigation of the ANN approximator ability to produce parameter values with an acceptable error at an increased range of model parameters. To illustrate this problem consider an exponential model:

$$y(t) = \theta_3 \left[\exp\left(\frac{-t}{10\theta_1}\right) - \exp\left(\frac{-t}{\theta_2}\right) \right] \tag{1}$$

where θ_1, θ_2 and θ_3 are 3 unknown parameters. The response (1) is quite ubiquitous, e.g. it may represent biological multicompartmental systems [9] or step response of electronic circuits [7]. It is assumed in this example that the parameters can take their values from a cuboid as follows:

$$\{\Theta:\ 0.6 \le \theta_1 \le 1.4,\ \ 0.2 \le \theta_2 \le 1.0,\ \ 0.8 \le \theta_3 \le 1.2\} \tag{2}$$

A well-known single-hidden layer perceptron, with sigmoidal nonlinearity in the hidden layer and linear combiner at the output, was trained using a two-stage procedure [10]. Three separate networks have been trained, one for each parameter. The inputs to the networks were the model observations, i.e. values of *y(t)* calculated from (1) at 3 distinct time moments: t_1=0.56 ms, t_2=2.61 ms and t_3=12.9 ms. The training patterns were calculated from (1) on a regular lattice of *8* points per each parameter range, total of *512* points. The respective parameter values were the target quantities to be learned by the ANNs. After the training, the networks were tested on a denser grid of *20* points per parameter, total of *8000* test vectors.

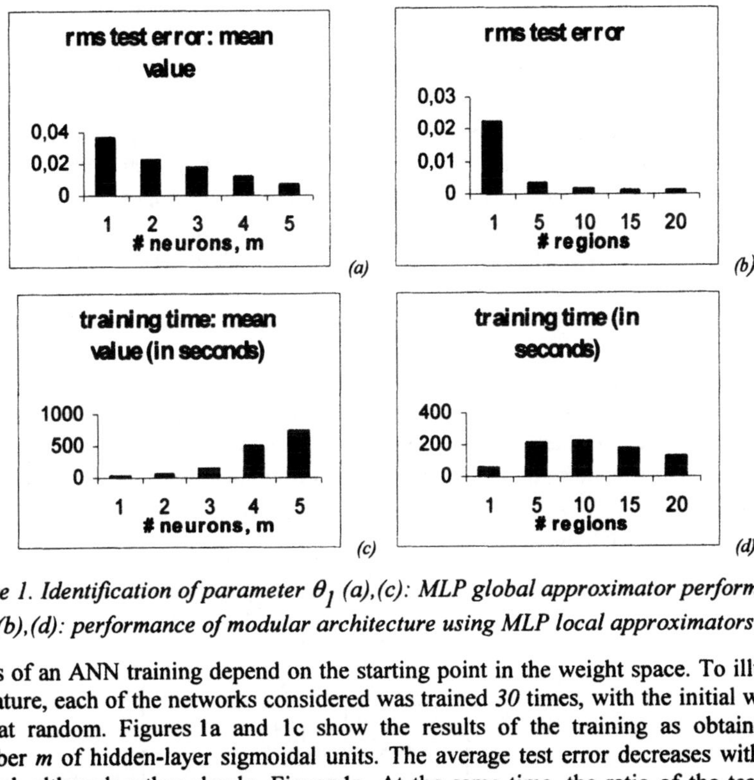

Figure 1. Identification of parameter θ_1 (a),(c): MLP global approximator performance (b),(d): performance of modular architecture using MLP local approximators

Results of an ANN training depend on the starting point in the weight space. To illustrate this feature, each of the networks considered was trained *30* times, with the initial weights taken at random. Figures 1a and 1c show the results of the training as obtained for a number *m* of hidden-layer sigmoidal units. The average test error decreases with *m* as intended, although rather slowly, Figure 1a. At the same time, the ratio of the test error standard deviation to the error mean value <u>increases</u> with *m* (not shown in Figure 1). Thus by increasing the network complexity in order to achieve higher approximation accuracy, one introduces more uncertainty to the training process instead. (This can be explained by the fact that with the increased number of weights of the ANN, there appear new local minima of the error function which is minimised during the training. It becomes then more likely that the training algorithm is trapped by a local minimum and is unable to reach the global one.) An increase of *m* leads to an exponential increase of the training time, Figure 1c. Similar results were obtained in case of the other parameters of the model (1). In conclusion, this approach to achieving higher accuracy of approximation (through an increase of the number of hidden-layer neurones) does not seem practical. There is a need for developing an ANN architecture whose training will produce more predictable results.

3. MODULAR ANN ARCHITECTURE

It is well known, that in case of polynomial approximation of a smooth single-variable function $v(x)$ on a compact set of real numbers, $x \in \langle a,b \rangle \subset \Re$, the demand for higher-order polynomial terms, necessary to maintain a given accuracy level, increases with the range $r=|b-a|$ of the independent variable. For a narrow range r, retaining a linear term only may prove satisfactory. Similar property is attributed to single- and multi-variable function expansion using other base functions, not only polynomials. This suggests a solution to the problem addressed in the previous section. Namely, it is proposed to split the observation domain Φ of the mapping into a number q of non-overlapping regions Φ_i, $i=1,...,q$ such that their union covers the whole domain, $\Phi_1 \cup \Phi_2 \cup ... \Phi_q = \Phi$. Since a unique continuous mapping from the observation space to the parameter space exists, there are q regions Θ_i in the parameter space, such that $\Theta_1 \cup \Theta_2 \cup ... \Theta_q = \Theta$, each corresponding to a respective region Φ_i, $i=1,...,q$. As q increases, the range of parameters within each of the regions Θ_i and the range of observations within each of the regions Φ_i both decrease. The approximation task within each observation region becomes then simpler, since the mapping $\theta = f^{-1}(\varphi)$, $\varphi \in \Phi_i$ is closer to a linear relationship compared to its complexity over the whole domain, $\varphi \in \Phi$, where the nonlinearity is more pronounced. As a consequence, lower complexity approximator network is needed within each of the regions to maintain a given accuracy level. It is necessary, however, to decide which region a given observation vector belongs to.

The idea discussed can be realised using a modular ANN architecture which employs a classifier and a bank of q approximators [11]. Each approximator is preoptimised in terms of its weight vector $\boldsymbol{w}$ to provide lowest possible error within the respective region. The classifier first allocates an acquired observation vector to one of the q regions, $\varphi \in \Phi_i$. The proper approximator ANN is invoked next, to produce the parameter vector values within the region Θ_i, $\hat{\theta} = g_i(\varphi) \cong f^{-1}(\varphi) \in \Theta_i$. The block diagram of the proposed modular architecture is shown in Figure 2.

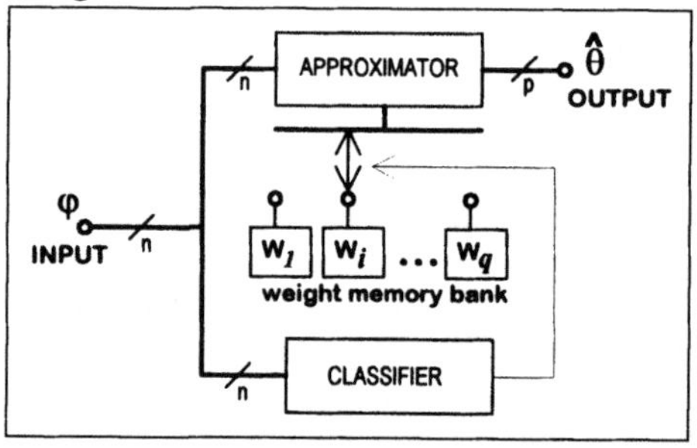

Figure 2. Proposed modular ANN architecture

Computer simulation was performed to apply the modular ANN to the identification of parameters of model (1). A minimum-Euclidean-distance classifier was trained to tessellate the observation domain into q = *5, 10, 15* and *20* regions. Two approximator structures were then taken into account, an MLP with 2 sigmoidal neurones (*11* weights) and a rational function (RF) network [12] with *13* weights. The q approximators were trained using a numerical minimisation routine and then tested on a dense grid of *8000* points covering the whole parameter domain (2). Some of the results obtained are shown in Figure 1, for q MLPs approximating parameter θ_1. The standard deviation of the RMS error over a number of training sessions tends to decrease with q from already low values, so it is not considered here. With increasing q, the RMS test error decreases to much lower values than those obtained for the single MLP, compare Figure 1a and Figure 1b. The total time needed to train a lower-error modular ANN structure tends even to decrease with the error value, Figure 1d. On the contrary, the single-approximator architecture needs longer training times to achieve lower errors, Figure 1c.

These results are encouraging and prove the usefulness of the modular ANN architecture. In general, for a given *q*, lower error values and shorter training times were experienced with the RF network which is a promising approximator architecture for this application. (It needs further investigation to decide which approximator structure, MLP or RF, is more suitable for the task of model parameter identification.) Training of the modular architecture is a better controlled and predictable process compared to the training of a single, large-size global approximator network. The price paid for that is a slightly increased time needed for identification, namely this time is extended by a period necessary to classify the observation vector. Parallel classifiers may help reduce this period to acceptable values.

4. MODULAR ANN HARDWARE SYSTEM

Functional design of the proposed mixed-signal prototype ANN parameter identification system, is illustrated in Figure 3. Control circuit, programmed by a PC host computer, initialises data sampling and controls data classification stage. First, signal generator excited by the control circuit delivers a test voltage waveform to the system under test (SUT). It is assumed that the observed SUT is characterised by known parametric model and its time response to the test signal depends on model parameters. Control circuit determines time moments of SUT response sampling also. The sampling circuit delivers the vector of SUT response samples to the input of the neural network approximator. Simultaneously, the same signals are fed to the analog classifier, which takes decision as to which weight set should be used for the parameter approximation stage. Classifier output is sampled by the host PC which loads appropriate weight vector to the approximator network. Finally, SUT parameters are estimated by the ANN approximator.

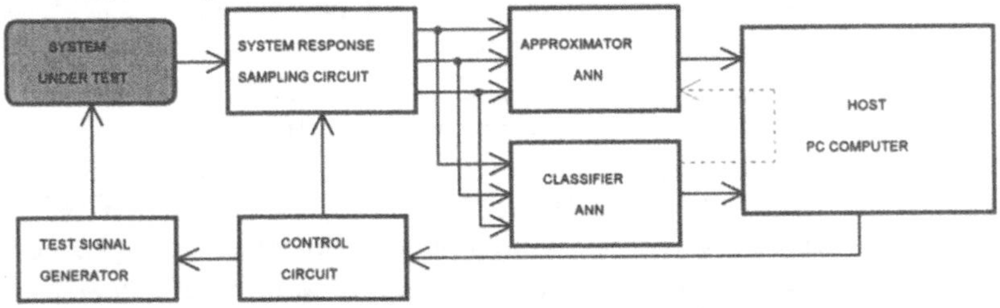

Figure 3. Block diagram of ANN-based parameter identification system

System parameter identification process is controlled and supervised by the host PC computer. The host PC controls time moments of SUT response sampling, initialises SUT observation and performs some data processing. It reads also A/D converted signal samples, classifier internal voltages and its output. It downloads approximator network architecture and appropriate weight vector to signal processor extension card, transfers SUT observations to the approximator network and reads estimated SUT parameters. Learning processes of the classifier and approximator ANNs are also controlled by the PC host computer. Detailed description of ANN-based parameter identification system can be found in [13].

5. DISCUSSION AND CONCLUSIONS

The prototype design of ANN system for parameter identification was presented in this paper. The role and operation of its functional blocks were discussed. The system is now under CAD design and simulation tests. Once have been built, the prototype ANN mixed-signal system will serve as a tool for experimental verification of the novel, recently proposed technique for fast and robust model parameter identification [4]. Extensive tests are planned using a variety of system under test, including parametric testing of CMOS circuits based on power supply transient responses [7]. The effect of observation noise on the identification process accuracy will also be experimentally investigated to verify results of numerical simulation [6], [7]. It is expected that a significant advance to the measurement and instrumentation technology will be made by introducing the proposed ANN-based parameter identification technique into practice.

REFERENCES

[1] J. Beck and K. Arnold, *Parameter estimation in engineering and science*, Wiley, 1977.

[2] A. Atkinson and A. Donev, *Optimum experiment design*, Oxford Science, 1992.

[3] K. Hornik, *Some new results on neural network approximation*, Neural Networks, vol. 6, 1993, 1069-1072.

[4] A. Materka, *Application of neural networks for dynamic system parameter estimation*, 14th Int. Conf. IEEE Eng. Med. Biol. Soc., Paris, 1992, 1042-1045.

[5] J. Bandler and Q-J. Zhang, *Optimization techniques for modeling, diagnosis and tuning*, in Analog Methods for Computer-Aided Circuit Analysis and Diagnosis, T. Ozawa (Ed.), Marcel Dekker, 1988.

[6] A. Materka and S. Mizushina, *Parametric signal restoration using artificial neural networks*, IEEE Trans. Biomed. Eng., vol. 43, no. 4, April 1996, 357-372.

[7] A. Materka and M. Strzelecki, *Parametric testing of mixed-signal circuits by ANN processing of transient responses*, J. of Electr. Testing: Theory and Applications, no. 9, 1996, 187-202.

[8] A. Annaswamy and S-H. Yu, *θ-adaptive neural networks: a new approach to parameter estimation*, IEEE Trans. Neural Networks, vol. 7, no. 4, 1996, 907-918.

[9] A. Bahil, *Bio-engineering*, Prentice -Hall, 1981.

[10] A. Materka, *Application of artificial neural networks to parameter estimation of dynamical system*", IEEE Conf. Meas. Instrum, Hamamatsu, Japan, 1994, pp. 123-126.

[11] A. Materka, *Classifier-approximator modular neural network for accurate estimation of dynamic system parameters*, Appl. Math. And Comp. Sci., vol. 6, no. 3, pp. 447-461.

[12] H. Leung, S. Haykin, *Rational function neural network*, Neural Comput., **5**, 1993, pp. 928-938.

[13] P. Pelczynski, M. Strzelecki and A. Materka, *Design of Mixed-Signal ANN Prototype System for Model Parameter Identification*, 4th Int. Conf. MIXDES '97, Poznan, 463-468.

16

MIXED A/D VLSI ARCHITECTURE FOR THE EMULATION OF NEURO-FUZZY MODELS

Juan Manuel Moreno, Jordi Madrenas, Spartacus Gomáriz and Joan Cabestany

Department of Electronic Engineering
Technical University of Catalunya
Building C4, c/Gran Capitŕ s/n, 08034 - Barcelona
SPAIN

ABSTRACT

In this paper we shall present a novel analog hardware architecture which is able to implement efficiently neuro-fuzzy models. By combining the main features of digital and analog alternatives it will be possible to provide a high degree of flexibility (in terms of number of inputs, number of membership functions per input and number of fuzzy rules) when handling real world tasks. The performance estimations obtained for the proposed architecture show that it yields a good area/throughput ratio, thus making it suitable for a wide range of applications.

1. INTRODUCTION

Since its inception by Zadeh in the 60s [1], fuzzy set theory has gained acceptance in a wide range of applications, encompassing control, pattern recognition, signal processing, etc. Furthermore, in order to meet the constraints imposed by real world tasks, several hardware implementations have been proposed, including digital [2] as well as analog [3] alternatives. However, until now no architecture has been proposed capable of combining the advantages of both alternatives (i.e., the easy reconfigurability of the digital solutions and the compactness typical of the analog implementations), thus allowing for a high degree of flexibility (in terms of number of inputs, number of membership functions per input and number of rules to be included in the inference engine) when handling real world tasks.

In this paper we shall propose a novel analog systolic architecture which is able to combine the main features of digital and analog solutions, providing in this way an efficient alternative for the implementation of fuzzy models. Furthermore, by using some of the organization principles of an existing mixed analog-digital architecture [4], the resulting architecture will allow for the emulation of a wide range of artificial neural network models.

We shall consider for our architecture the Sugeno fuzzy model [5], which states that the crisp output provided by the fuzzy system, o, is given by:

$$o = \frac{\sum_{i=1}^{P} \omega_i \cdot p_i}{\sum_{i=1}^{P} \omega_i} \quad , \quad p_i = \sum_{k=0}^{n} x_k \cdot c_k \tag{1}$$

where:

x_i: Crisp inputs to the system
n: Number of inputs
P: Number of fuzzy rules included in the fuzzy inference engine
ω_i: Value resulting from evaluating the ith fuzzy rule
c_k: Proportionality coefficient for the kth crisp input

In the next section we shall outline the main features of the proposed VLSI architecture which will permit the emulation of fuzzy models based on the principles stated previously.

2. PROPOSED ARCHITECTURE

The general organization of the proposed architecture is depicted in Figure 1. As it can be seen, the architecture can be divided in five main building blocks. The *inference* block is in charge of evaluating the fuzzy rules (obtained as a *minimum* combination over the fuzzy values provided by determining the *m* fuzzy membership functions defined for each crisp input), while the *consequent* block weights these fuzzy rules with the linear combination of the inputs with the corresponding coefficients. The outputs provided by both blocks are given as analog currents. The blocks labelled *S()* perform an addition of their input currents, being the block *D()* in charge of yielding the final output, *o*, by calculating the division indicated in expression (1). The main function of the analog memory blocks is to store the partial results given by the aggregation of consequents and fuzzy rules performed by the *S()* blocks.

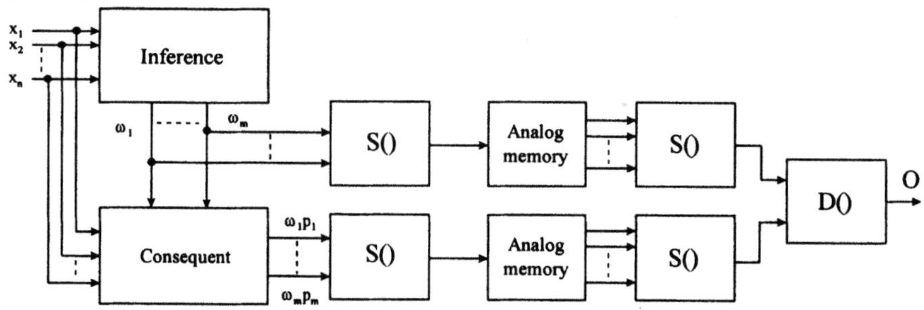

Figure 1. General organization of the proposed architecture

The basic principle used to emulate the Sugeno fuzzy model consists of obtaining at each emulation step as many fuzzy rules as membership functions are defined for each input variable. For this purpose, the *inference* block is organized as depicted in Figure 2.

In this figure we have represented an example of the configuration for the *inference* block in the case the fuzzy system has to cope with 3 inputs (x_1, x_2, x_3) and there are 2 membership functions defined for each input. The blocks labelled $F_{k,xi}$ in the figure implement the kth membership function for the ith input, while the blocks labelled *Min()* provide at their outputs the minimum of their two input currents. As it can be deduced, the membership functions for the two first inputs are calculated sequentially, while the membership functions for the remaining input are calculated in parallel, so that at each emulation step two fuzzy rules comprising the 3 inputs are yielded. As a consequence, the 8 rules of the system are calculated in 4 emulation cycles. In a general case of a fuzzy system which has n inputs and m membership functions defined per input, the inference block would provide the complete set of fuzzy rules in m^{n-1} cycles.

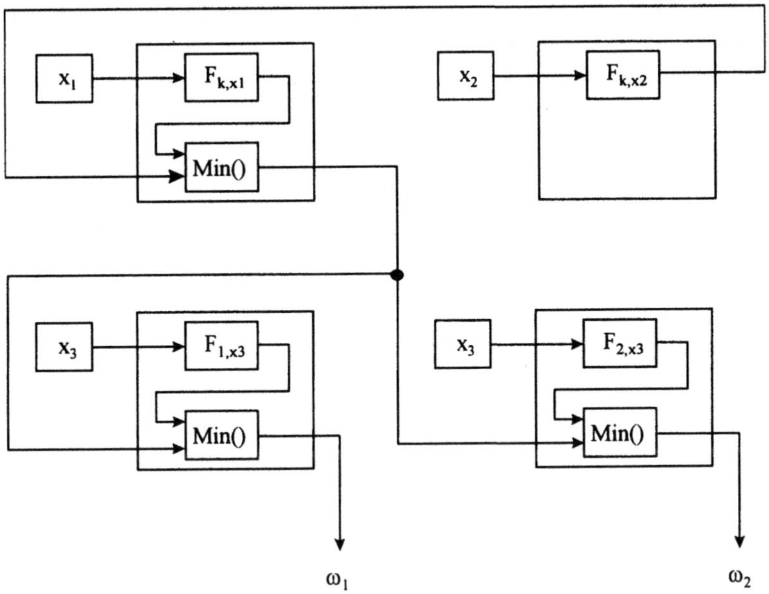

Figure 2. Organization of the inference block

The organization of the *consequent* block for the same example explained previously is depicted in Figure 3. The blocks labelled $m_1()$ in this figure are analog-digital multipliers, while the blocks labelled $m_2()$ are analog multipliers. The terms c_{ij} are the weight coefficients for the ith input at the jth emulation step, and can be stored in a digital memory. Since the structure used in this functional block is quite similar as that proposed in [4], it will be possible, with almost no control overhead, to use the proposed architecture to emulate also a wide range of neural models.

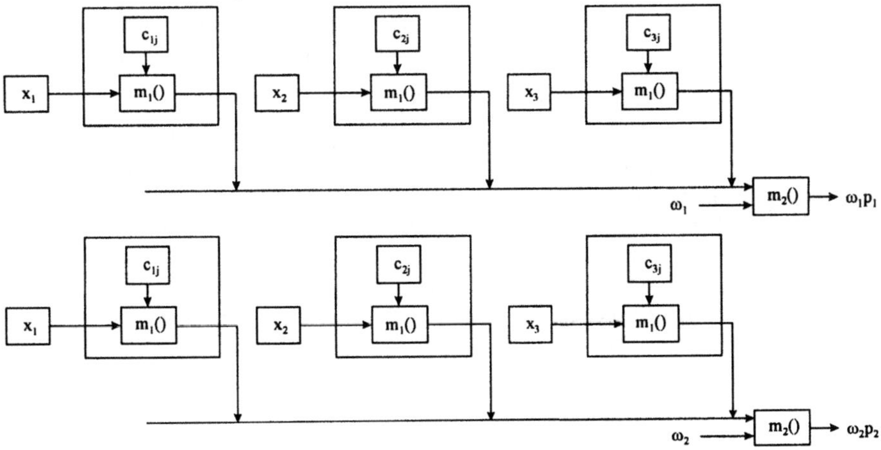

Figure 3. Organization of the consequent block

The organization and data flow proposed for the *inference* and *consequent* blocks is highly modular, allowing for the flexible emulation of different configurations for the fuzzy system. For instance, it can be demonstrated that a *inference* block composed of 6 basic cells and a *consequent* block integrated by 10 $m_1()$ blocks and 5 $m_2()$ blocks permit the emulation of the configurations stated in Table 1 (parametrized in terms of number of inputs, n, and number of membership functions per input, m).

Table 1. Possible configurations for a fuzzy system to be implemented with the proposed architecture

n	m
2	2
2	3
2	4
2	5

n	m
3	2
3	3
4	2

3. BASIC BUILDING BLOCKS

Regarding the *inference* block, two main functions have to be implemented: the membership function and the *Min()* function. Provided that the inputs to the system are given in voltage mode, a possible efficient implementation for the programmable membership functions could be that presented in [6], which is based on two coupled differential stages. The main advantage of this cell is given by the fact that its output is provided as an analog current, thus being able to be interfaced directly to the *Min()* function cell which can be derived from the *Max()* function proposed in [7].

For the *consequent* block, two main functions are required: the analog-digital and the analog multipliers. The analog-digital multiplication can be performed either by means of the basic cell proposed in [4] or just by a weighted array of current mirrors if the precision to be used is not larger than 8 bits. On the other hand, the analog multiplication may be provided by a Gilbert-like cell, similar to what has been recently proposed in [8].

The analog memory cells included in the proposed architecture can be implemented by means of regulated cascode current copier cells [9]. The characterization we have performed for these cells in a 1.2 μm CMOS technology shows that they are able to provide about 8 bit accuracy, which is enough for a wide range of applications.

Finally, the current addition blocks can be performed simply with an array of current mirrors, while the current division cell may follow the structure proposed in [10].

4. PERFORMANCE ESTIMATION

As it was indicated in section 2, the *inference* block of the proposed architecture is able to produce at each emulation cycle as many fuzzy rules as membership functions are defined for each input. In this way, if we consider a general fuzzy system with n inputs and m membership functions per input, the emulation of the whole system is completed in m^{n-1} cycles. Therefore, since the analog-digital multiplication can be performed in parallel with the membership function evaluation, the worst case (obtained for the emulation of a 4-input, 2 membership functions per input fuzzy system) total execution delay, t_d, for the proposed architecture can be estimated by the following expression:

$$t_d = m^{n-1} \cdot (t_f + 3 \cdot t_{min} + t_{mul} + t_{add} + t_{mem}) + t_{add} + t_{div} \tag{2}$$

where:

t_f: time required to evaluate a membership function
t_{min}: time required to evaluate a *Min()* function
t_{mul}: time required to evaluate an analog multiplication
t_{add}: time required to evaluate a current addition

t_{mem}: settling time of the analog memory cell

t_{div}: time required to evaluate a current division

Furthermore, since the basic cycle time is given by $(t_f + 3 \cdot t_{min} + t_{mul} + t_{add} + t_{mem})$, a quite conservative estimation for a 0.8 μm CMOS technology provides a maximum frequency for the system between 1 and 5 MHz, thus outperforming recent developments [3].

5. CONCLUSIONS

In this paper we have addressed the hardware implementation of neural as well as fuzzy models. By combining the solutions provided by analog and digital alternatives, we have proposed a novel analog systolic architecture whose sequencing scheme, together with the modular organization of its building blocks, permits a high degree of flexibility for emulating fuzzy models. Furthermore, the structural choice made for some of its functional blocks permits also to emulate a wide range of artificial neural network models with almost no control and area overhead.

The compact cells which integrate the main functional blocks of this architecture allow for a small area system realization, thus making the architecture suitable for flexible low cost applications. Finally, the concurrent execution scheme imposed in the data flow permits to attain a high processing speed, being therefore possible to handle real time applications.

Our current work is devoted to the characterization of the basic cells which constitute the architecture for a 0.8 μm CMOS analog technology. Furthermore, a precision analysis is being performed in order to determine the programming scheme to be used for the membership function generator cells.

REFERENCES

[1] L. A. Zadeh, *Fuzzy Sets*, Information and Control, 8, pps. 338-353, 1965.

[2] M.J. Patyra, J.L. Grantner and K. Koster, *Digital Fuzzy Logic Controller: Design and Implementation*, IEEE Transactions on Fuzzy Systems, Vol. 4, No. 4, pp.439-459, November 1996.

[3] S. Guo and L. Peters, H. Surmann, *Design and Application of an Analog Fuzzy Logic Controller*, IEEE Transactions on Fuzzy Systems, Vol. 4, No. 4, pp. 429-438, November 1996.

[4] J.M. Moreno, F. Castillo, J. Cabestany and J. Madrenas, A. Napieralski, *An Analog Systolic Neural Processing Architecture*, IEEE Micro, Vol. 14, No. 3, pp. 51-59, June 1994.

[5] T. Takagi and M. Sugeno, *Fuzzy Idendification of Systems and its Application to Modeling and Control*, IEEE Transactions on Systems, Man and Cybernetics, Vol. 15, pp. 116-132, 1985.

[6] M. Sasaki, N. Ishikawa, F. Ueno and T. Inoue, *Current-mode Analog Fuzzy Hardware with Voltage Input Interface and Normalization Locked Loop*, Proceedings of the IEEE International Conference on Fuzzy Systems, pp. 451-457, San Diego, 1992.

[7] J. Lazzaro et al., *Winner-Take-All Networks of O(n) Complexity*, Advances in Neural Information Processing Systems, Vol. 1, D.S. Touretzky (ed.), pp. 703-711, Morgan Kaufmann, 1989.

[8] D. Coué, G. Wilson and *A Four-Quadrant Subthreshold Mode Multiplier for Analog Neural-Network Applications*, IEEE Transactions on Neural Networks, Vol. 7, No. 5, pp. 1212-1219, September 1996.

[9] D. Macq, *Application des Copieurs de Courant dans les Circuits Analogiques CMOS*, Ph. D. Thesis, Université Catholique de Louvain, February, 1994.

[10] K. Bult and H. Wallinga, *A Class of Analog CMOS Circuits Based on the Sqare-Law Characteristic of an MOS Transistor in Saturation*, IEEE Journal of Solid-State Circuits, Vol. 22, No. 3, pp. 357-365, June 1987.

17

A NEW APPROACH FOR FINDING OPTIMUM DESIGN OF ELECTROSTATIC MICROMOTORS

Abdul Wahab A. Salman, Andrzej Napieralski, Marek Turowski and Grzegorz Jabłoński

Department of Microelectronics and Computer Science
Technical University of Łódź
al. Politechniki 11, 93-590 Łódź
POLAND

ABSTRACT

This paper presents a new approach for finding optimum configuration and geometric size of very small variable capacitance side drive micromotors in order to obtain the motor with the highest average output torque and minimum torque ripple. Letting all design parameters vary independently various torque curve characteristics can be calculated for different combinations of pole configuration. Modified circuit model is used for calculation of drive torque, which includes the influence of side planes of both stator and rotor poles, which leads to fast and more accurate estimations. The design process is automatically set up after choosing initial design parameters including initial geometric size.

1. INTRODUCTION

In recent years microsystems are becoming more and more popular. They contain standard electronic circuits, microsensors and microactuators fabricated on single silicon die. The principle of operation of micromotors used in microsystems can be either electrostatic or electromagnetic. The torques produced from the change of electrostatic field energy for smaller size rotary micromachines becomes larger, than that produced by electromagnetic field [1,2,6]. Therefore, various micromotors fabrication and design procedures are considered. The most important and successful micromotors are variable-capacitance (VC), side drive motors, due to the relative simplicity of design and fabrication [3,4,5,8]. Numerous practical applications are investigated recently with such type of micromotors, mainly in medical instrumentation, optical systems, microrobotics, and aerospace technology [4,5,7].

The scope of this work is to describe the method for finding optimal design of polysilicon, rotary, side drive electrostatic micromotors. The design procedure presented in this paper is used for finding the optimal geometric parameters that lead to highest average torque with the minimum torque ripple for a given pole configuration. Two types of pole configurations are considered with different ratios of stator pole number to rotor pole number, in order to find an optimal pole configuration to provide superior average torque and minimum torque ripple.

Simple modified circuit model is used, in order to simulate a large number of designs for different pole configurations. In this model, simple type of calculations are employed, taking into account the influence of side planes of stator and rotor poles for the estimation of equivalent capacitance. In order to examine the accuracy of proposed model, the results gained from our model were compared with two-dimensional results for actual calculations of drive torque [4,6,8].

2. DESIGN PROBLEM

2.1. Selection of Initial Parameters of Micromotor.

This section will illustrate how the relevant parameters are selected. In actual design, the first parameter selected is the rotor-stator polysilicon layer thickness. It sets a limit on the maximum rotor diameter and minimum air-gap spacing. Decreasing rotor thickness or increasing rotor diameter may lead to rotor warpage due to residual stresses in polysilicon. The standard process uses a 2.2 μm thick polysilicon layer for typical fabricated rotor radii of 50-65 μm, and minimum air-gap spacing of 1.5-2 μm, which are easily attainable. During optimization procedure, motor drive torque can be normalized for one micrometer of axial motor length, in order to keep this parameter constant. The micromotor drive torque scales linearly with rotor-stator polysilicon layer thickness. Another most critical parameter affecting drive torque is the air-gap spacing, which is inversely affecting drive torque. However, the reduction of air-gap spacing is not only limited by the electric field breakdown in air gap spacing, but also limited by the micromotor fabrication process.

Drive torque may be also increased by increasing rotor radius, since it scales approximately linearly with the radius of rotor. The initial value of the motor radius is selected to satisfy the performance requirement on the maximum drive torque and maximum operating speed of micromotor. It must be noted that by increasing the radius of the motor the upper limit of the number of pole/phase is increased. However, the upper limit of the number of pole/phase is determined by the minimum pole width beyond which the corresponding maximum drive torque is degraded by increasing the number of poles [2,4,7]. Therefore, for a given rotor radius, and air gap spacing, the optimal geometric size of micromotor will be determined by appropriate selection of the stator and rotor poles width, that will be discussed in the following section.

2.2. Finding the Optimum Geometry of Electrostatic Micromotor.

The choice of optimal geometric size of micromotor will base on maximum drive torque (or speed), minimum torque ripple, and ease of fabrication, in order to yield the highest probability of successful motor operation and obtaining a high operation speed.

For any pole configuration, the geometry of micromotor can by defined by a number of independent parameters. The number of parameters can be reduced to five, leading to the following variables (See Figure 1):

- Rotor radius r_2.
- Air-gap spacing δ.
- Stator pole width / stator pole pitch ratio $t_1\% = t_1 / t_{p1}$.
- Rotor pole width / rotor pole pitch ratio $t_2\% = t_2 / t_{p2}$.
- Slot radius /rotor radius ratio $r_{slot}\% = r_{slot} / r_2$.

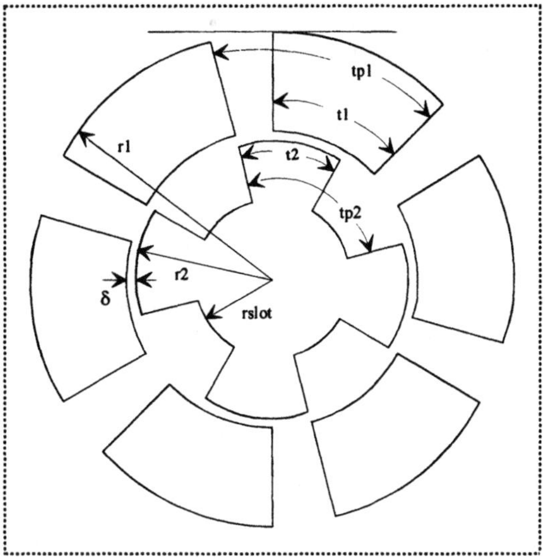

Figure 1. Top cross-sectional view of micromotor (with linear dimensions).

For a given pole configuration, the optimal geometric size of micromotor is found by varying the parameters $t_1\%$, $t_2\%$, and $r_{slot}\%$, respectively (i.e. , how much the pole pitch of the stator and rotor are occupied by the stator electrode and rotor teeth and how deep the rotor teeth penetrate in the rotor). Therefore, the optimal micromotor design is determined by means of a successive sampling of design space.

2.3. Optimal Pole Configuration.

The design procedure will be also used to examine different types of pole configuration, to select the pole configuration providing the highest average torque. The optimal pole configuration is aimed to further increasing drive torque (and hence the average torque), where drive torque is enhanced by increasing the number of poles per phase. However, for a given geometric size of motor, there is a limit on pole configuration beyond which the corresponding drive torque is degraded by increasing the number of poles. Two types of pole configuration are considered with pole number ratio 12/4, and 12/8. Comparison is made in order to select the best motor which has the optimal pole configuration and optimal geometric size.

3. TORQUE CALCULATION

The optimal design of an electrostatic VC micromotor starts from the evaluation of motive torque. Where the output drive torque is given by the derivative of co-energy of the electrostatic field with respect to the angular displacement of rotor position [6,7].

$$T(\theta) = \frac{\partial W(\theta)}{\partial \theta} \propto \frac{\partial C}{\partial \theta} \qquad (1)$$

Applying a modified simple circuit method the torque versus rotor position characteristic can be calculated. Such method not only treats the rotor-stator pole face as parallel plate capacitor [6], but also takes into account the effects of side planes of both rotor and stator poles. Both sides of stator and rotor are contributed by an additional capacitance. Such additional capacitance can be approximately estimated by the formula of the parallel plate capacitor, and it depends on the angular position of the rotor, pole pitch ratios of stator and rotor, and slot radius to rotor radius ratio. Therefore, the equivalent capacitance C (for each rotor position step of 3°) across the driving electrodes results from the parallel connection of elementary capacitances between stator and rotor in the radial direction.

Output torque characteristics for different dimensions and pole configurations are computed, in order to find an optimal geometric shape of the motor. Therefore, a large number of calculations must be performed. By varying all geometrical parameters, respectively, large number of motor models can be created. Output drive torque versus rotor position characteristic can be calculated from the torque produced by 8 different relative positions (i.e. during one half electrical period of rotor starting from aligned position to misaligned position). Table 1 shows the comparison between torque values computed by simple modified parallel plate approximation and values simulated by two dimensional method [6] as a function of rotor position for typical 3:2 side drive motor with 12 stator and 8 rotor poles.

Table 1. Comparison of torque values computed by simple modified method with values simulated by two dimensional method.

Rotor position angle (degree)	Torque simple modified method (pNm)	Torque two dimensional method (pNm)
0°	0	0
3°	5.71	5.6
6°	8.12	7.5
9°	8.55	8.3
12°	8.69	8.5
15°	8.53	8.2
18°	8.12	7.6
21°	2.96	2.6

As it can be seen from Table 1, the modified simple method gives a good approximation of torque values over the whole torque curve versus rotor position. Figure 2 shows optimal output drive torque characteristics for typical micromotor, of 12/8 pole configuration, 50μm rotor radius, 2μm air-gap spacing, and 2.2μm thickness. The average torque T_{av} can be calculated by integration of the area under the torque characteristic. The torque ripple is defined as follows :

$$T_{rp} = \frac{T_{\max} - T_{\min}}{T_{av}} \tag{2}$$

T_{max} = maximum positive torque.

T_{min} = minimum positive torque.

4. RESULTS AND CONCLUSIONS

The typical polysilicon micromotor has 50μm rotor radius, 2μm air-gap spacing and 2.2μm thickness. The optimal geometric size is determined during the motor design to have minimum torque ripple. There are several designs that will give the same torque ripple, but comparison is made between these designs to select the motor with the highest average torque.

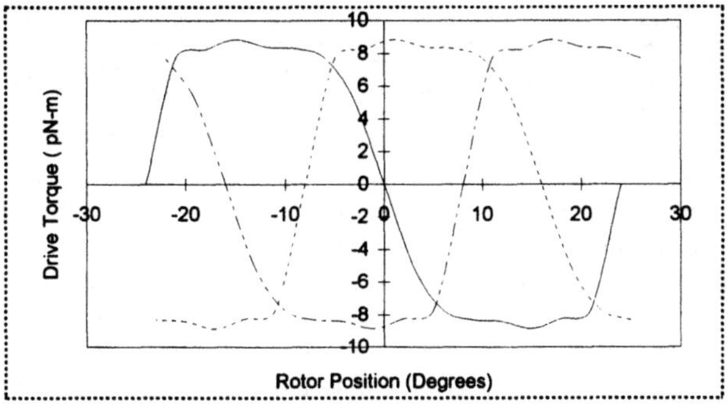

Figure 2. Optimal drive torque characteristics for 12/8 pole configuration micromotor of rotor diameter 100 μm.

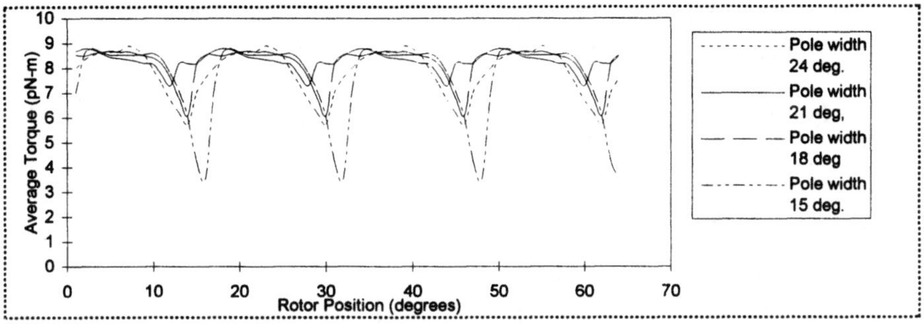

Figure 3. Output torque characteristics for micromotors of pole configuration12/8 designed to have different pole pitch ratios.

Figure 3 shows the output torque characteristics for 12/8 pole configuration micromotor designed to have different geometric sizes (i.e. different pole pitch ratios). From this curve, it can be observed that the optimal geometric shape of 100μm rotor diameter is obtained with the pole width of 21 degrees. Figure 4 also shows the output torque characteristics for two types of pole configurations of 12/8 and 12/4 micromotors designed to have the same rotor diameter and aspect ratio. From Table 2 and Figure 4, it can be seen that the 12/8 (i.e. 3:2) pole configuration will provide superior average torque and less torque ripple, in comparison with 12/4 pole configuration. Finally, Figures 5, and 6 show the outlines and dimensions of these two optimal designs.

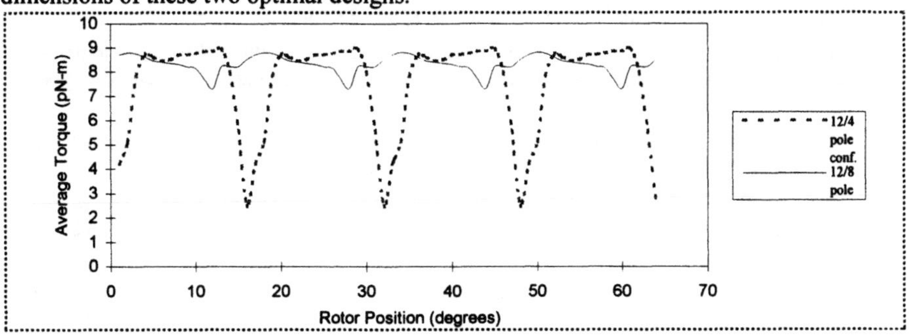

Figure 4. Output torque characteristics for micromotors of 12/4, and 12/8 pole configuration

Table 2. Optimal torque results for different pole configuration micromotors.

Pole Configuration	Average Torque (pN-m)	Torque Ripple (%)
12/4 pole	7.74	85%
12/8 pole	8.3	18%

As it can be seen from this optimization procedure, all the initial and final design parameters have been properly selected and determined to meet the performance requirements on maximum drive torque and minimum torque ripples, and to satisfy all the constraints imposed by the limitations of microfabrication process. Since the micromotor is electrically linear system, the optimal values of rotor radius and air-gap spacing can by easily determined. Therefore, for a given radius and air-gap spacing, an optimal geometric shape can be determined by appropriate selection of stator and rotor pole width parameters. Finally, optimal pole configuration can be found by comparing different types of pole configurations.

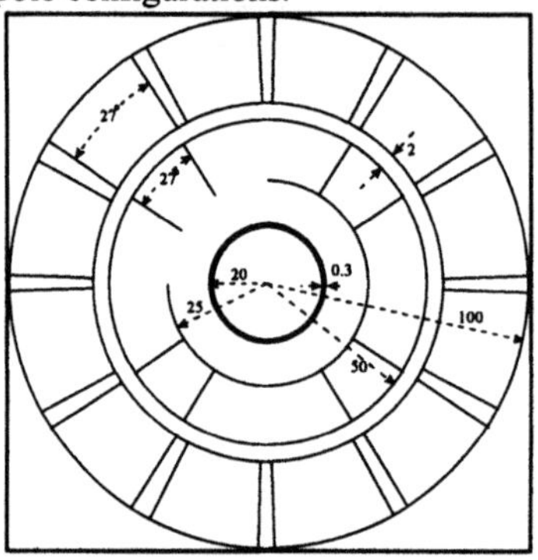

Figure 5.Optimal geometric size of 12/4 micromotor(all dimension in μm).

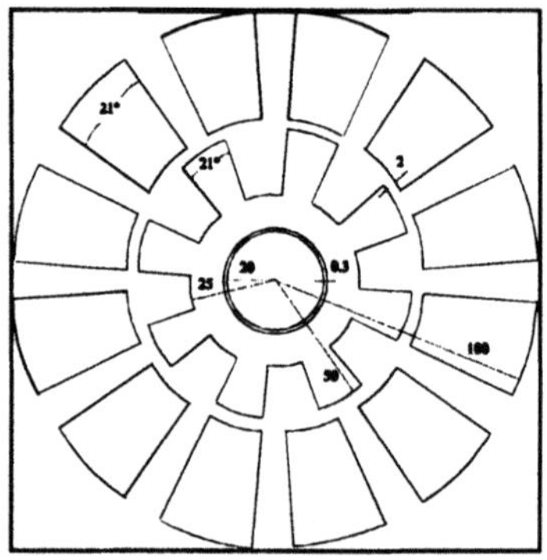

Figure 6.Optimal geometric size of 12/8 Micromotor (all dimension in μm)

REFERENCES

[1] W. S. N. Trimmer and K. J. Gabriel, *Design considerations for a practical electrostatic micromotor*, Sensors and Actuators, 11, 2, 189-206, March 1987.

[2] S. F. Bart, T. A. Lober, R. T. Howe, J. H. Lang and M. F. Schlecht, *Design considerations for microfabricated electric actuators*, Sensors and Actuators, 14, 3, 269-292 , July 1988.

[3] J. H. Lang and S. F. Bart, *Toward the design of successful electric micromotors*, IEEE Solid-State Sensor and Actuator Workshop, June, 1988, pp.127-130.

[4] M. Mehregany, S. F. Bart, L. S. Tavrow, J. H. Lang and S. D. Senturia, *Principles in design and microfabrication of variable-capacitance side-drive motors*, J. Vac. Sci. Technical, Vol. A8 , pp.3614-3624, July/Aug. 1990.

[5] M. Mehregany, S. D. Senturia and J. H. Lang, *Micromotor fabrication*, IEEE Trans. Electron Devices, Vol. 39, No. 9, pp.2060-2069, September 1992.

[6] Di Barba P., Savini A. and Wiak S., *2-D Numerical simulation of electrostatic micromotor torque*, Second Int. IEE Conf. On Computation in Electromagnetics, Nottingham, 12-14 April 1994, pp.227-230

[7] T. B. Johansson, M. Van Dessel, R. Belmans, and W. Geysen, *Technique for finding the optimal geometry of electrostatic micromotors*, IEEE Trans. Industry Applications, Vol. 30, No. 4, July/Aug 1994.

[8] Turowski M., Jabłoński G., Napieralski A and, Wiak S., *Electromechanical and Thermal Modelling of IC-processed Micromotors*, 2nd Workshop "Mixed Design of VLSI Circuits" - MIXDES'95, Kraków, Poland, 29-30 May 1995, pp. 343-348.

18

ONE-CYCLE CONTROLLED BOOST CONVERTER FOR MICROSYSTEMS

Noureddine Senouci, Francis Therez and Daniel Esteve

Laboratoire d'Analyse et d'Architecture des Systèmes du Centre National de la Recherche Scientifique LAAS/CNRS 07, Avenue Colonel Roche 31077 Toulouse Cedex4 FRANCE

ABSTRACT

A DC-DC Boost converter for the purpose of biomedical microsystem applications is presented. The input voltage is provided by an array of six photovoltaics cells integrated on silicon and permitting a supply voltage of 3.5V. To control the output voltage we have used a One-Cycle Control technique which is applied for the first time on Boost converter. The use of this technique is supported by its inherent benefits in terms of sturdiness, automatic correction of switch errors and, especially, rejection of line perturbation which, in our case, is due to the power changes at the photovoltaic cells output. The converter presents a simulated conversion efficiency of 78.4 percent which can be enhanced by decreasing the output voltage. Control of the power DMOS is carried out at 500kHz, the different stages of the feedback control being achieved in full-custom with a 1.2µm P-well CMOS process over a total surface area of 0.3mm^2 and supplied by a pump charge circuit integrated into the same chip and providing an output voltage of 5V with a current of 1mA. A behavioural simulation of the control loop was first achieved with the ELDO simulator prior to the electrical simulation. The theoretical analysis together with the electrical implementation of the one cycle control technique are described.

1. INTRODUCTION

In a rapidly growing number of system applications, the high integration level of microelectronics alone is not sufficient. As a results, over the last years, several developments from various groups around the world have been gathered together as a multidisciplinary research field with the aim of introducing a new system integration including micromechanical elements, actuators, sensors, and signal processing into a same system, referred as Microsystem. They are based on the use of IC fabrication methods and more particularly make use of silicon as the integrated substrate . However, prior to the development of these microsystems, many problems remain unsolved including microfabrication and packaging, reliability, miniature power supply, interfacing, advanced information processing, control etc. [1].

Thus we have focused on a miniature power supply and investigated the various possibilities of utilizing a 3.5 voltage from photovoltaic cells integrated on silicon. This supply in expected to provide a voltage of 5V to bias electronic circuits of the microsystem as well as high voltage values to control actuators like an electrostatic micropump [2,3]. The converter solution with serial-parallel capacitors-switch cells operating in accordance with the charge pump mode or MARX generator [4,5] has been envisaged. This solution presents drawbacks however like the dependence of the output voltage on the number of cells and the small output current. Another alternative based on DC-DC converters has also been considered resulting in the fabrication of a DC-DC Boost converter.

2. BEHAVIOURAL SIMULATION OF THE ONE-CYCLE CONTROLLED BOOST CONVERTER

The One-Cycle Control technique is a nonlinear method of control effective on the duty-ratio of a switch so that, in each cycle, the average value of a switched converter variable is equal or proportional to the control reference in the steady-state or transient state [6]. This technique takes advantage of the pulsed and nonlinear nature of switching converters and achieves instantaneous control of the average value of the chopped voltage or current.

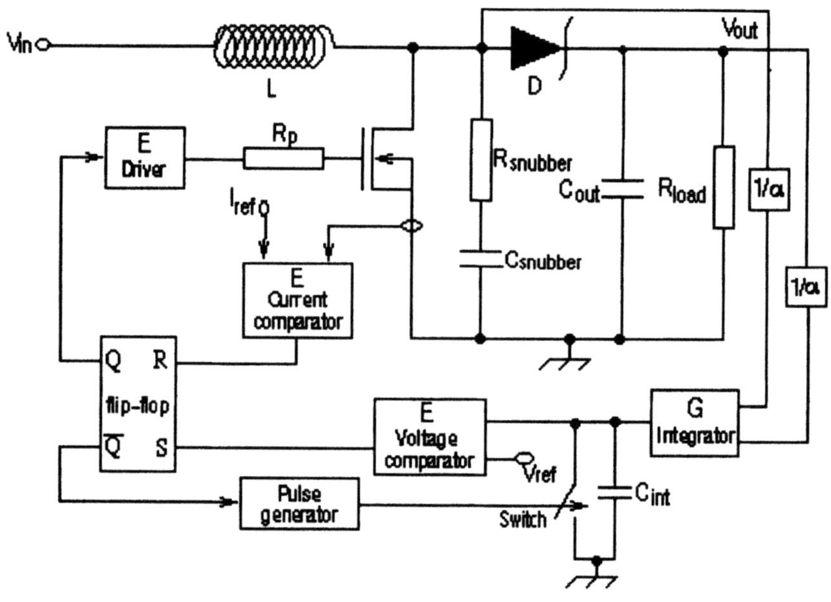

Figure 1. Behavioural modelling of the Boost converter

The behavioural modelling is an extension of the controlled source of the Exx and Gxx type (SPICE and ELDO simulators). It allows a simple description of all electronic functions by its transfer function expressed either literally or by means of table of values, and this, both in the temporal (VALUE, TABLE) and frequency fields (LAPLACE, FREQ, CHEBYSHEV).

First, we use it to verify circuit operation and determine the electrical characteristics of the different blocks. Figure 1 gives the Boost schematic, with its control loop. Used for the behavioural simulation and Figure 2 shows the simulation results.

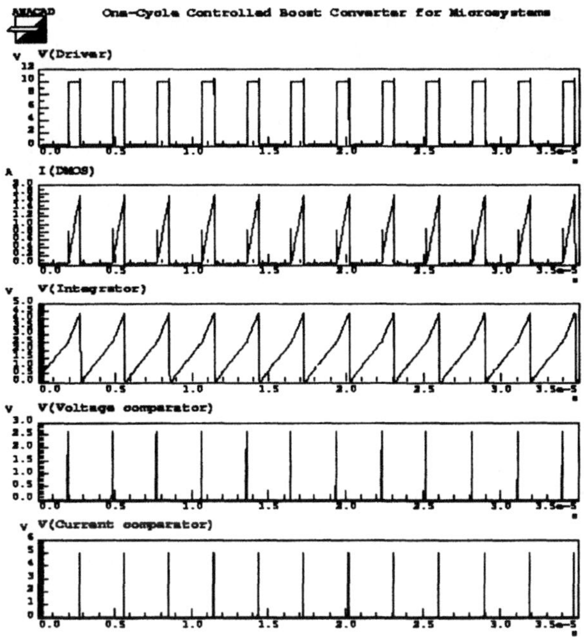

Figure 2. Main chronographs of the Boost converter for a behavioural simulation

3. ELECTRICAL IMPLEMENTATION OF THE ONE-CYCLE CONTROL TECHNIQUE

The behavioural simulation allowed verification of the circuit operation and determination of the electrical parameters of the different stages. In this section, we will present the electrical circuit of the control loop and a comparison between the results obtained from behavioural and electrical simulations.

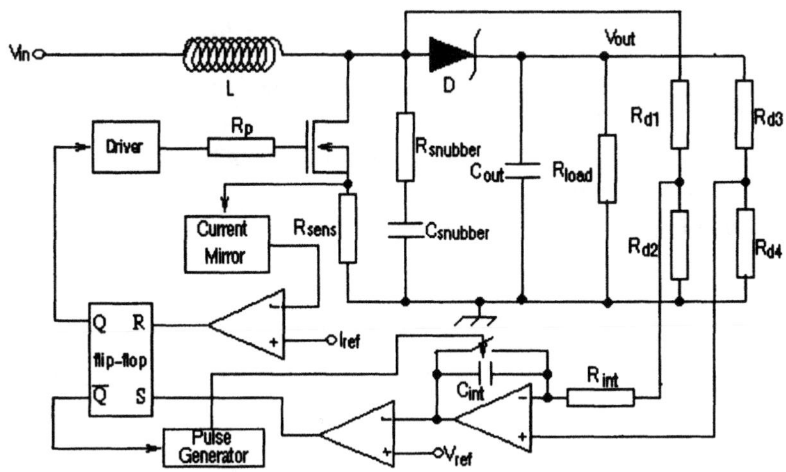

Figure 3. Electrical implementation of the circuit of Figure 1

Conduction of the DMOS switch for Vint≥0 sets the charge coil to a maximum current I_M. Detection of this current causes DMOS clamping and transfer of the energy to the load during time T_d where the diode is "ON" (Figure 4).

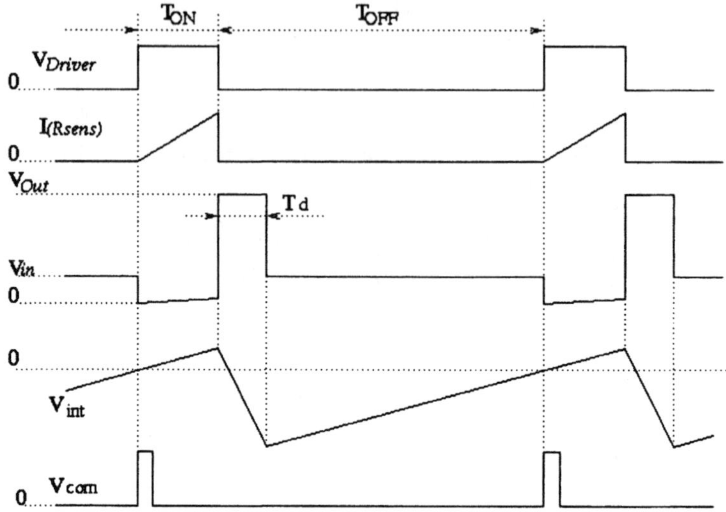

Figure 4. Theoretical waveform of the One-Cycle Controlled Boost converter

- from 0 to T_{ON}, DMOS is on and charges the coil to a maximum current I_M corresponding to induction limit below magnetic material saturation.

$$I_M = \frac{V_{in}}{L} \cdot T_{ON} \tag{1}$$

- From T_{ON} to $T_{ON}+T_d$, the energy $LI^2/2$ is transferred to the output capacitor and to the load through the diode D.

$$I_M = \frac{V_{out} - V_{in}}{L} \cdot T_d \cong \frac{V_{out}}{L} \cdot T_d \tag{2}$$

During time T_d, the diode voltage V_D is applied to the integrator input and is equal to the difference of input and output voltage.

- At T_s=T_{ON}+T_{OFF} the output comparator changes and triggers DMOS when V_{int}=0

For the integrator, the following is obtained:

- from 0 to T_{ON} : (V_{DC} is the cathode diode voltage)

$$V_{\text{int}} = \frac{1}{RC} \int_0^t V_{DC} dt \tag{3}$$

- from T_{ON} to $T_{ON}+T_d$ (V_{DA} is the anode diode voltage = drain voltage of DMOS)

$$V_{\text{int}} = \frac{1}{RC} \int_0^t V_{DC} dt - \frac{1}{RC} \int_{T_{ON}}^{T_{ON}+T_d} V_{DA} dt \tag{4}$$

- from T_{ON}+ T_d to T_s=T_{ON}+T_{OFF}

$$V_{\text{int}} = \frac{1}{RC} \int_{o}^{t} V_{DC} dt \tag{5}$$

Therefore, the control law is:

$$\frac{1}{T_S} \int_{o}^{\delta T_S} V_D dt = V_{ref} \tag{6}$$

And the output voltage becomes:

$$V_{out} = \frac{T_S}{T_{ON}} \cdot V_{ref} \tag{7}$$

Figure 5 shows the electric circuit chronographs from Figure 3 simulated with the ELDO simulator. These signals are in good agreement with the theoretical signals described in Figure 4 and those obtained from the behavioural simulation of Figure 2.

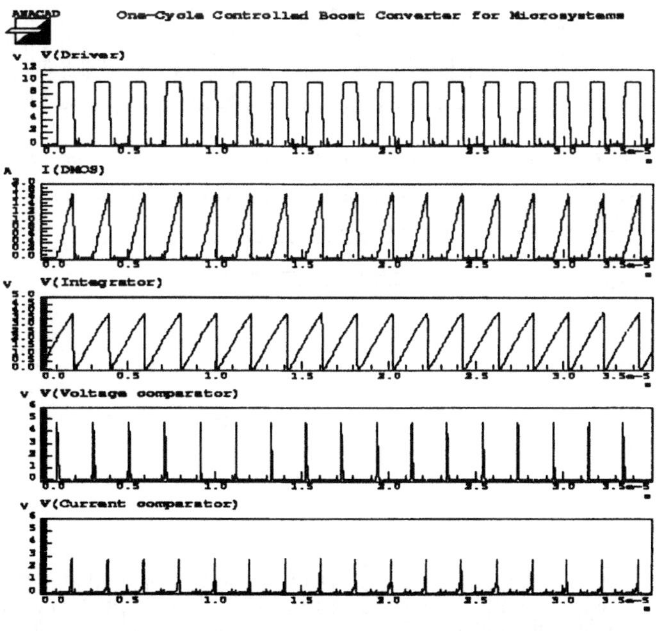

Figure 5. Main waveforms of the Boost converter for a electrical simulation

4. REJECTION OF THE PERTURBATION AT THE INPUT

As stated above, the output voltage of one-cycle control technique is proportional to the reference voltage and independent of the input voltage, so that if a perturbation signal is applied to the converter input, this technique will rejects it, owing to the integration of the changing diode-voltage in real time which causes the integration of the slope variation of their integrated diode-voltage and variation of the duty-ratio "δ".

Figure 6 shows a simulation result, in which a perturbation signal was applied to the converter input for a behavioural simulation, and Figure 7 shows the same results obtained by electrical simulation. Clearly, both simulations are in good agreement and the output is not affected by the input change.

Figure 8 shows the layout of the circuit of Figure 2 obtained by using the "Magic" software.

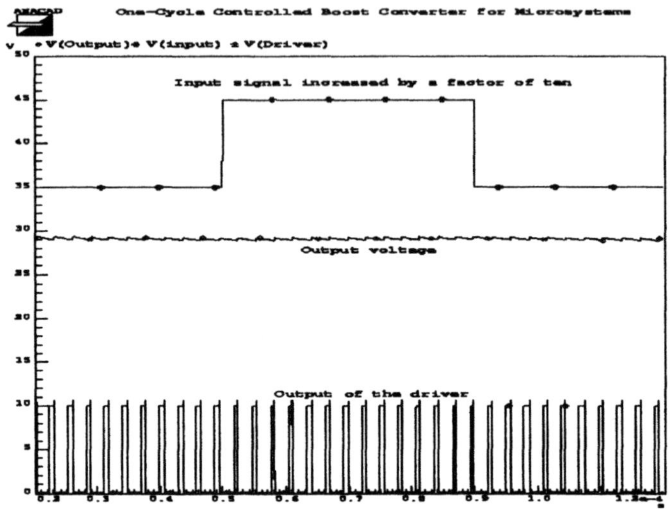

Figure 6. Rejection of the input perturbation for the behavioural simulation

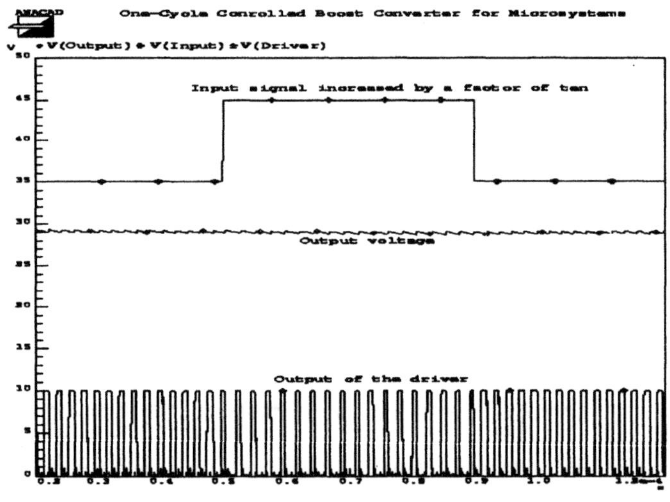

Figure 7. Rejection of the input perturbation for the electrical simulation

5. CONCLUSION

A solution for the supply of the microsystem for biomedical application has been based on the provided use of DC-DC converter units. The first stage is a Boost converter which receives at its input a voltage provided by an array of six photovoltaic cells integrated on silicon. To reject the perturbations at the converter input, due to the variation of the incident light, we used a control technique which, amongst many advantages, features the possibility to reject the input perturbations.

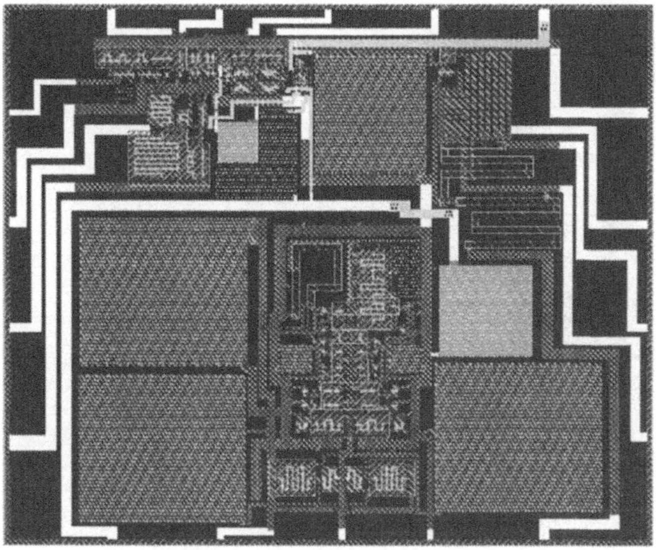

Figure 8. Layout of the control loop

We have presented the circuit for a behavioural simulation as well as the electrical circuit implementing the one-cycle control technique. The proposed circuit presents an conversion efficiency of 78.4 percent, which can be enhanced by decreasing the output voltage. The simulation results show that the circuit can reject the input perturbations. The control loop circuit dissipates 4.94mW and occupies a $0.3mm^2$ area.

REFERENCES

[1] S. Fatikow, U. Rembold and G. Wöhlke, *A Survey of the Present State of the Art of Microsystem technology*, First workshop on Micro-Robotics and Systems. Karlsruhe, pp.1-19, 15-16 June 1993

[2] D. Bosch, B. Heimhofer, G. Mück, H. Seidel, U. Thumser and W. Welser, *A silicon microvalve with combined electromagnetic/electrostatic actuation*, SENSORS AND ACTUATORS, A,
37-38, pp.684-692, 1993

[3] R. Zengerbe, W. Geiger, Richter, J. Ulrich, S. Kluge and A. Richter, *Application of Micro Diagram pump in Microfluid Systems*, 4th International Conference on New Actuators, ACTUATORS94, Bremen, Germany, June 15-17, 1994

[4] A. Lamantia, P.G. Maranesi and L. Radrizzani, *Small-Signal Model of the Cockroft-Walton Voltage Multiplier*, IEEE Trans. ON POWER ELECTRONICS, Vol.9, No.1, January 1994

[5] Kwa-Sur Tam and Eric Bloodworth, *Automated Topological Generation and Analysis of Voltage Multiplier Circuits*, IEEE Trans. ON CIRCUITS AND SYSTEMS, Vol.9, No.1, Junuary 1994

[6] Keyue M. Smedley and Slobodan Cuk, *One-Cycle Control of Switching Converters*, IEEE Trans. ON POWER ELECTRONICS, Vol.10, No.6, pp.625-633, November 1995

PART IV

Design Methodologies

19

DESIGN FOR REUSE: HDL BASED GRAPHIC DESIGN ENTRY FOR PARAMETRIZABLE AND CONFIGURABLE MODULES (A CASE STUDY)

Anna Boszko

Institute of Electron Technology
Al. Lotników 32/46, PL-02-668 Warszawa
POLAND

ABSTRACT

The paper describes different aspects of VHDL based design methodology proposed for creating a set of parametrizable and configurable modules for multimedia applications using (V)HDL based Graphic Design Entry Tools. Synthesizable VHDL code for designed parts has been considered for designing and for modeling purposes as well. Usage of Graphic Design Entry Tools for design management, configuring tool, mixed level simulation and automatic generation of synthesizable HDL code has been discussed.

1. INTRODUCTION

Synthesizable VHDL code of a designed component can be approved as a high-level portable model. It has at least one feature of reusability: it is technology portable. Portability over technology libraries has been delivered and assured by the foundries supporting several synthesis tools. Portability over design methodologies - e.g. considering diverse "input dialects" of synthesis tools - this is the challenge for designers. How "high" the abstraction level of the current synthesizable VHDL description is, depends strictly on the applied synthesis tool.

Another aspect in this context is the need of maintaining the feature reusability of every designed module over time - facing continuously changing standards, emerging requirements and evolutioning CAD environment.

The reported configuration is **Synopsys VSS**, **Synopsys DC** v.3.5a and **Cadence DFWII 9502**. For Graphic Design Entry Tool and automatic (V)HDL code generation stands **speedCHART** Rel.3.4.0 - a commercial tool, which has not been supported for education purposes by EUROPRACTICE consortium.

2. DESIGN MANAGEMENT

The described project was performed in academic research and development environment - quite inhomogeneous and changing one. Not all designer team members attend the project phases from the beginning till the end. The project manager introduced some common rules to all present code pieces (models); that means introducing of changes and testing their impact. At the same time every newly created design part had to follow them. Many rules proposed by [5] were applied as well. Basing partially on some pieces VHDL code which have already been written, the design specification was created in form of text document with several pictures, tables, block and timing diagrams. This specification was growing continuously all the time because of two main reasons: while designing, more and more details were discovered and design refinements had to be undertaken, respectively whole parts rewritten; the second reason - the described approach allows generally introduction of quick idea changes, which is a very attractive but dangerous feature, hardly to resist for a particular perfectionist designer.

3. CHARACTERISTICS OF THE APPROACH

According to the proposed by [2] classification, the following two groups of prototyping approach features can be extracted:

- As ***explorative prototyping characteristics*** of the approach

  Requirement or specification errors were recognised and corrected, lack of specification refined. The dynamics of the whole system was modeled and tested, while using certain timing generic values inside of non synthesizable parts of the design (technology specific RAM models, delivered by the foundry). The language and communication problems (misunderstandings) between the team members were subsequently recognised and eliminated. The working version of the synthesizable model became a validated last version of the specification.

- As ***experimental prototyping characteristics*** of the approach

  Refinement from the very abstract level (idea described by a written document) of description to a level closer to the implementation, which can be checked every moment and for every part for synthesis effectiveness. Checking the interaction between the individual functional blocks. The design and implementation requirements could finally be examined - for the resulting netlist representation's time behaviour normally differs very slightly from the real structure. The result of the experimental prototyping could be considered as a part of component library, supporting the system and component design.

As a result - a consistent and complete interface specification of the system components. Disadvantage of this mixed approach here are obviously additional effort and costs involved in comparison to the case if the specification were firm and complete from the beginning. This causes delays and changes according to the planned project schedule.

So - why not apply an old concept of reusability, doing as much as possible only once?

4. REUSABILITY REQUIREMENTS

The main requirements, after [6], that a component should meet in order to be reusable are:

- ***Interoperability:*** the component should be available for different CAD tools.
- ***Documentation:*** a key factor for reusability. A set of documents should contain: design document being the real specification of the module, reference guide, data sheet, application note, known problems and solutions.
- ***Technical support and maintenance:*** crucial aspects for the final commercialisation of the component set.
- ***Proven quality:*** depending on quality criteria.
- ***Configuration tools:*** for components which are modifiable in their functionality.
- ***Validation tools:*** helping the user in the simulation of the developed system.

- ***Error detection support:*** complex components should have methods to detect incorrect operation by the driving units.
- ***Manufacturing test support:*** in form of a set of test vectors or a BIST circuitry to do the test. This is specially important in the case of configurable components, for changing functionality with each particular configuration. Not so difficult for FPGA implementations.

It will be demonstrated how this quite unique (V)HDL based Graphical Design Entry Tool can meet most of these requirements while delivering concurrently: interoperable (V)HDL sythesizable code, design documents which are easy readable real design specifications, and facilitate the configuration, validation together with error detection and testing support as well.

5. DESIGN EXAMPLE

The developed structure contains a MIMD based motion estimation processor for MPEG2 encoding. It is expected to be able to perform motion estimation on a large search area with different user programmable algorithms [4]. The main characteristics here are the presence of two identical processing modules, which operate in parallel. Each module consists of a 24 bit RISC controller with its own instruction RAM, a pixel processor with extended input word size and a register file with sorting capability. The chip is intended to communicate with an external SDRAM and video input and output buffers by means of 16 bit bus interface ruled by a small micro-programmable controller with own micro-code RAM.

On the other side it is operating under control of a Host CPU which loads the necessary information into the appropriate places and takes the results. The last functionality (together with initialization and debugging mode of the whole structure) is being ensured by means of a Host Interface unit, controlling the access to the Host Interface Bus and the 256 16-bit words General Purpose RAM as well. It is out of the scope of this paper to describe the function of the whole chip in detail; it should be mentioned though that the structure contains as much hardware as possible. The limit is a reasonable chip area.

In the following sections we will see the presentation of the proposed design approach and design flow applied to a small part of the architecture - namely the Host Interface and Host Interface Bus control. As described above it is the Host Interface unit, which enables hierarchical controlling function of the external host CPU. This can serve for the following functions: ***Initialization*** of all configuration registers and writing program code into the processor module's instruction memories during starting phase as well; ***Loading*** information during operating phase into the configuration registers and modules, writing picture rates information into the appropriate buffers and weight matrix into the GP RAM at the picture rate; ***Debugging,*** e.g. reading all addressable registers (including the processing modules) connected to the Host Interface Bus for monitoring the chip at any time.

These three tasks can be fulfilled additionally in two different operating modes (potentially two pin programmable versions of the chip): - the ***fast mode*** using a 16 bit data input/output and 10 bit address input (one chip register can be read or written in one clock cycle); - the ***slow mode*** using only 8 bit address/data input/output external pins, multiplexing them so that one chip register can be read or written in four cycles; additionally a '***burst mode***' for address part is provided, where the starting address is incremented automatically, resulting in $(2n + n)$ cycles reading or writing of n 16 bit data words. The Host Interface controls the Host Interface Bus access by means of arbiter logic unit, which selects the source/target of the written/read information (requests come from either the external CPU or one of the processing units on chip). One of them is the GP RAM (see Figure 3).

From the point of view of our proposed design methodology, two parts of the Host Interface block are interesting for further consideration: the ***arbiter logic*** controlling the bus access (which has to be configurable depending on the number of graphic processors on the same chip) and the ***timing controlling block*** for serial operating mode of the chip. Figure 1 shows the design document (coming from speedCHART State Diagram Editor) in form of state diagram of the timing logic. This document contains specification of the functionality, facility to extract (also graphically) the control part from the datapath section, so it allows a very efficient automatic generation of VHDL code resulting in more than three pages VHDL code applying the simplest coding style.

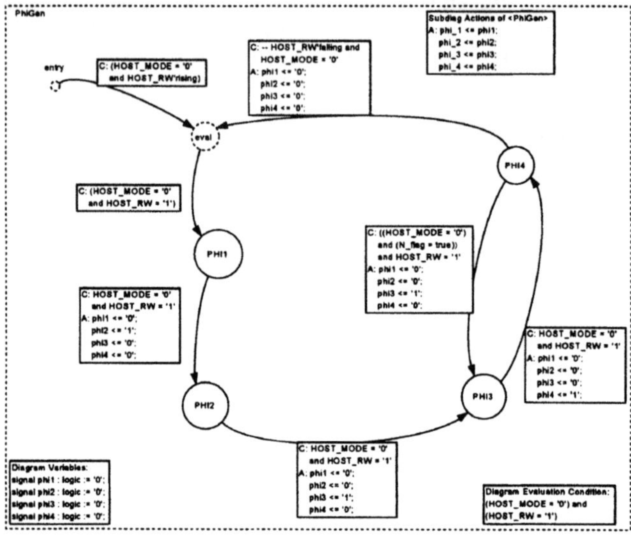

Figure 1. State Diagram document describing the function of timing control logic

While using automatic HDL code generator the designer can choose among different coding styles. One is not pushed anymore to write as much readable as possible [5], while typing the code manually - it doesn't matter if the resulting VHDL code is more or less readable for a human being - the only criterion is then obtaining the most efficient physical structure implementation from given composition of technology library vs synthesis tool. The related structures for the clock system from Figure 1 after synthesis using two different coding styles are shown on Figure 2a and 2b respectively. The structure 2a is a result of coding style expressing the state machine as procedure (task) only, while the structure 2b has been obtained from enriching the same description with option "FSM equations", which caused about 50 percent more description's VHDL code. The structure 2b. is better, considering the area and performance, even if obtained at identical synthesis directives and constraints resp.

Figure 3 documents the complete test bench for Host Interface unit design and simulation under speedCHART environment. It includes the GP RAM behavioural model and two very simplified behavioural models of processing units as signal sources (resp. targets). This can be treated as an example of using speedCHART as Design Management Tool. Instead of conditional "if then generate" VHDL construct, that has not been supported by the Tool, fetching or not diverse components or their functional modifications may replace this configuring feature.

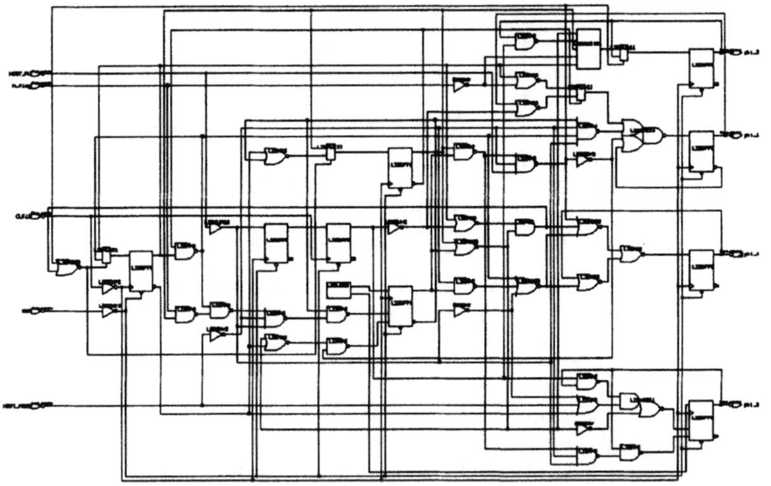

Figure 2a Resulting structure of the multiplexing clock system after synthesis

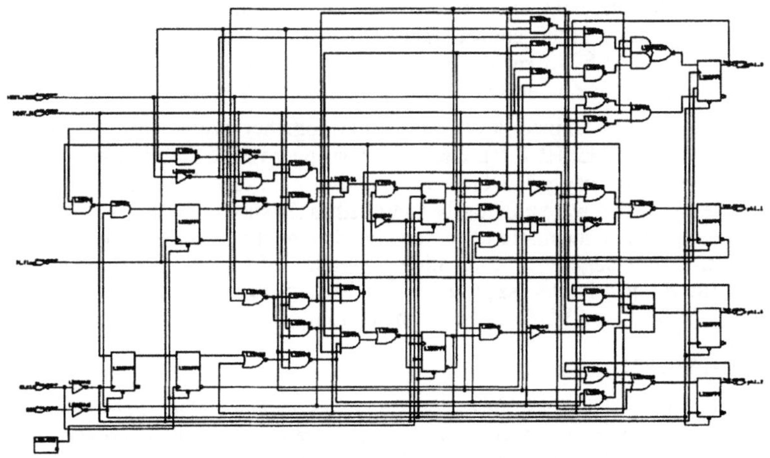

Figure 2b Second resulting structure of the multiplexing clock system after synthesis of VHDL architecture using additional FSM equation variables

It is visible, that parametrizing generic value for the word length is used - signed as $WIDTH keyword. Usage of generic width (respectively word length) was intended for creating potentially a component set parametrizable (for both: simulation and synthesis use) while applying evolutionary prototyping approach. Figure 3 does not only represent a 16 bit data and 10 bit address processor interface – as it is in fact by setting properly some generic parameters inside of the document. It has the meaning of N bit data and A bit address computer interface, which co-operates with an appropriately parametrized RAM unit and controls a compatible tri-state bus I/O interconnection. If we consider the pin-programmable function enabling parallel/serial operating mode, so we could not wish more reusability planes for this small design.

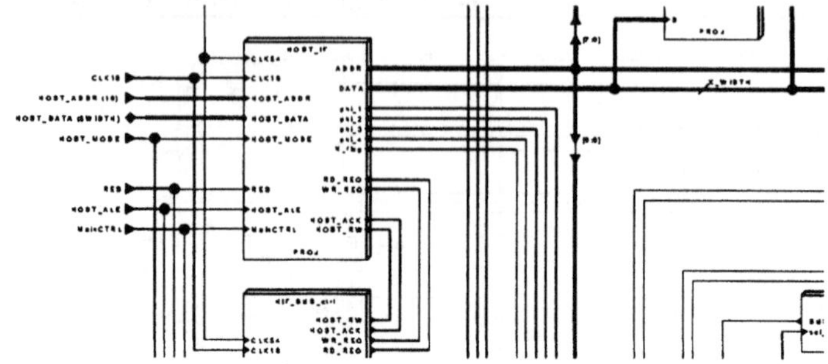

Figure 3. Test bench for the Host Interface using graphic editor as example of Design Management and Configuring Tool

6. CONCLUSIONS

Reusable components have to be parametrizable, e.g. in their generic bus width or word length, and configurable in their functionality. Presented Graphical Design Entry Tool enables creating such kind of components while documenting them in a very readable way, doing in fact the exact specification of the functionality. It enables scenario comparison ruling the synthesis tool while allowing application of different coding styles and choosing the maximal efficiency of resulting synthesized structures. It supports several synthesis tools, in this context assures the requirement of interoperability. The tool allows creating hierarchy of such components enabling their reuse in far more complex, hierarchical and incremental designs.

It is able to perform mixed level simulation together with some behavioural (not necessarily syntzesizable) HDL representations basing on RTL descriptions. Creating such kind of models could pay off if there is a need of accelerated simulation while reusing parts of a design which were previously approved as finished and reliable. From very clear and multiple used documentation and specification of the created design unit up to the unique feature of managing and evaluating generation of various target implementations - all these features justify its usage as Design Management Tool complementary to the convenient VHDL based VLSI IC design methodology.

REFERENCES

[1] W. Hobbs, Intel Corp. USA, *Model Availability, Portability and Accuracy - An IC Vendor's Perspective*, Proceedings of the Workshop on Libraries, Component Modeling and quality Assurance 26-29.04.1995, IRESTE - IHT, Nantes, France, pp 5 - 20.

[2] S. Olcoz, L. Ekterna, et al. *Prototyping: the Bottom Line of VHDL System Simulation*, Proceedings of the Workshop on Libraries, Component Modeling and quality Assurance 26-29.04.1995, IRESTE - IHT, Nantes, France, pp 21 - 38.

[3] V. Preis and S. März-Rössel, *Aspects of Modeling a Library of Complex and Highly Flexible Components in VHDL*, Proceedings of the Workshop on Libraries, Component Modeling and quality Assurance 26-29.04.1995, IRESTE - IHT, Nantes, France, pp 39 - 58.

[4] M. Gumm, et al., *A High Fault-Coverage Design-For-Testability Approach for a MIMD based Multimedia Processor*, accepted for European Design &Test Conference March '97, Paris, France.

[5] P. Sinander, *The Usage of VHDL in the European Space Agency*, Proceedings of the Workshop on Libraries, Component Modeling and quality Assurance 26-29.04.1995, IRESTE - IHT, Nantes, France, pp 141 - 152.

[6] Y. Torroja, T. Riesgo, et al., *Design for Reusability: Generic and Configurable Designs*, Proceedings on System Modeling and Code Reusability, VHDL User's Forum in Europe April 1997, Toledo, Spain (invited paper).

20

HIERARCHICAL TEST GENERATION FOR DIGITAL SYSTEMS

Marina Brik, Gert Jervan, Antti Markus, Jaan Raik and Raimund Ubar

Technical University of Tallinn
Ehitajate tee 5, EE0026, Tallinn
ESTONIA
Fax: (+372) 620 2253, e-mail: raiub@pld.ttu.ee

ABSTRACT

A hierarchical test generator for digital systems described on register-transfer (RTL) and gate levels is presented. The system is supposed to consist of control and data parts coupled with global feedback. The generator implements a novel test generation approach based on using multiple abstraction levels of alternative graph (AG) models. The uniform AG representation allows application of common modelling methods and procedures on all abstraction levels. Experimental results showing the efficiency of the approach are provided.

1. INTRODUCTION

Test generation for real life digital circuits on the gate-level is extremely complex. It has been shown that test generation for combinational circuits is an NP-complete problem [1]. Gate-level test generation for sequential circuits is even more complex and remains still an unsolved problem in practice. During recent times, as a possible solution, hierarchical test generation methods have evolved [2-5] which take advantage of higher abstraction levels (behavioural or register-transfer (RT) levels) information while generating tests for gate-level faults. The system is considered at different levels, and tests are created on these levels by separate tools. Both, top-down and bottom-up strategies are known. In the bottom-up approach, tests generated at the lower level will be later assembled at the higher level. Current paper considers the top-down approach, where constraints extracted at higher level [6] are considered when deriving tests for the lower level. In the approach discussed below, different design abstraction levels of the system are represented by *alternative graph* (AG) models [7,8]. This feature provides for a uniform model representation and an application of common procedures throughout the levels. As the result, the complexity of the problem can be reduced and the efficiency of the ATPG will be increased.

The paper is organized as follows. Section 2 explains the concept of AGs for representing digital systems at different abstraction levels, Section 3 describes the structure of the test generator, and in Section 4, experimental results are given.

2. THE MODEL

In this paper an approach based on AGs (or decision diagrams) for test generation is used. AGs serve as a mathematical basis for solving a wide spectrum of test tasks, resulting in a uniform fault model and a restricted set of standardized procedures. AGs were proposed the first time for test generation in [9]. Unlike the analogical binary decision diagrams (BDD) [10] introduced for representing Boolean functions, AGs describe both the functions and the structural features of a circuit or a system.

AGs can be regarded as a way of representing programs (procedures, algorithms), or as a data structure to be manipulated by programs, or as a way to concisely represent test knowledge that we have about the system. This universality of AGs makes it possible to transform different information we have about the system easily and directly to the form which is most suitable for solving test design or diagnosis tasks.AG-s describe digital systems on mixed logical and functional levels, which can include random logic, traditionally treated at the gate level, as well as digital systems like microprocessors, controllers etc., traditionally described at the procedural or RT levels. The fault model developed for AGs covers in a uniform way a wide class of faults represented at different levels like stuck-at faults, opens, shorts, functional faults [11], faults for VHDL descriptions [12] etc. The fault model defined on AGs can be regarded as a generalization of the classical gate-level stuck-at fault model [7].

Alternative graph is defined as a non-cyclic directed graph whose nodes are labelled by variables, constants or algebraic expressions. For each combination of values for node variables there exists always a corresponding activated path from the starting node to some terminal node. This relationship describes a mapping from a Cartesian product of the sets of values of all node variables to the joint set of values of labels in the terminal nodes. Therefore, by AGs it is possible to represent arbitrary digital functions $Y = F(X)$, where Y is the variable whose value will be calculated on the AG and X is the vector of all variables which belong to the labels of the nodes in the AG.

When using AGs to describe complex digital systems, we have, at the first step, to represent the system by a suitable set of interconnected components (combinational or sequential ones). At the second step, we have to describe these components by their corresponding functions which can be represented by AGs. AGs which describe digital systems at different levels may have special interpretations, properties and characteristics, however, the same formalism and the same algorithms for test and diagnosis purposes can be used, which is the main advantage of AGs.

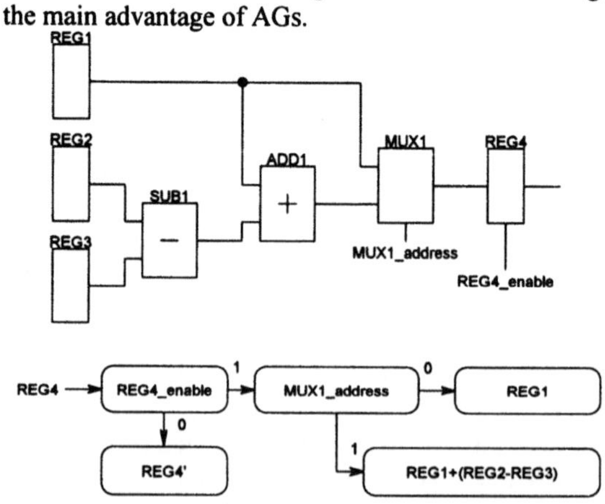

Figure 1. AG Representation of a Data Path

In Boolean level descriptions, the AG variables are Boolean (i.e. single bits), whereas in register-transfer level AG descriptions, in general case, multi-bit variables are used. RT-level descriptions in general are partitioned into control and datapath parts. The control part is described by an AG where non-terminal nodes represent current state and inputs for the control part, and terminal nodes are for representing the next state and control signals going to the datapath. The datapath can be described as a set of AGs in a way where for each register and each signal fanout an AG corresponds. Here, non-terminal nodes represent the control signals coming from the control part and terminal nodes represent signals of the datapath, i.e. primary inputs, registers, operations. Figure 1 shows a datapath fragment and its corresponding AG model.

3. DESCRIPTION OF THE TEST GENERATOR

The general structure of the test generator is presented in Figure 2. It consists of a hierarchical datapath test generator, a control part test generator, and a high-level AG model synthesizer. From RT-level VHDL description, high-level AG model is created, and from the gate-level netlists of high-level components, low-level AG-models are created. The AG-models will serve as an input for both test generators. Current system uses Design Compiler by Synopsys Inc. for the logic-level synthesis. The RT-level VHDL description and a VHDL library of FUs, containing generic bit-width behavioural descriptions of the FUs, serve as inputs for logic-level synthesis. The low-level AG generator creates AGs from EDIF 2.0.0 netlist descriptions.

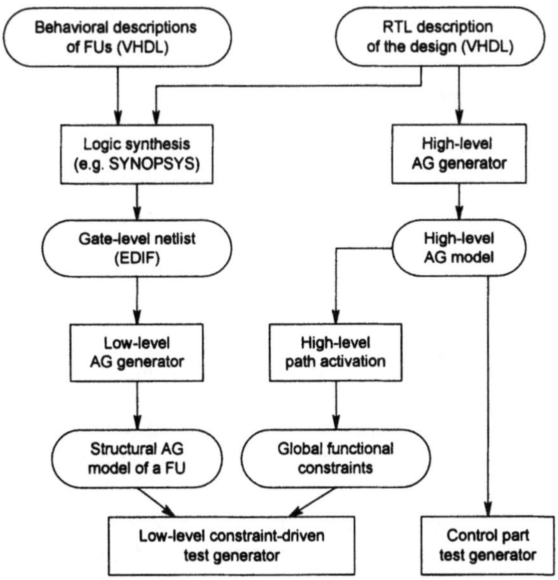

Figure 2. The AG-Based Test Synthesis System

The *control part test generator* works in the following way. Two types of faults will be considered: the faults caused by defects in the next state logic of the control part (transition faults), and the faults caused by defects in the control logic in the data part (output faults). Because of these two types of faults, the ATPG consists also of two main parts working together in the test generation system. The first of them, transition fault processor (TFP) begins to work from the initial state and will traverse step by step all the transitions of the finite state machine (FSM). In each step, the processor introduces all the faults activated at the current transition. Suppose, that N1 faulty machines and one fault-free machine were processed at the current step. The TFP passes all necessary inputs to the second part of the ATPG – to the output fault processor (OFP) for simulating the behaviour of the data part for all the activated N1+1 machines. The second task is to introduce now all the control faults in the data part activated at this transition. Suppose, that M1 additional faulty machines were produced in the data part at the current step. The OFP passes now all necessary inputs for all the N1+M1+1 machines back to the TFP for simulating the behaviour of the control part. In the same time, the OFP checks if some of the faulty machines will produce different output signals in the data part as expected. If a fault will be detected, the corresponding faulty machine will be removed from the list of simulated machines. The described procedure will continue for the next transition. Fault introducing continues until all not yet detected faults are considered. The procedure continues until all the faults will be detected. More details of the generator are given in [8].

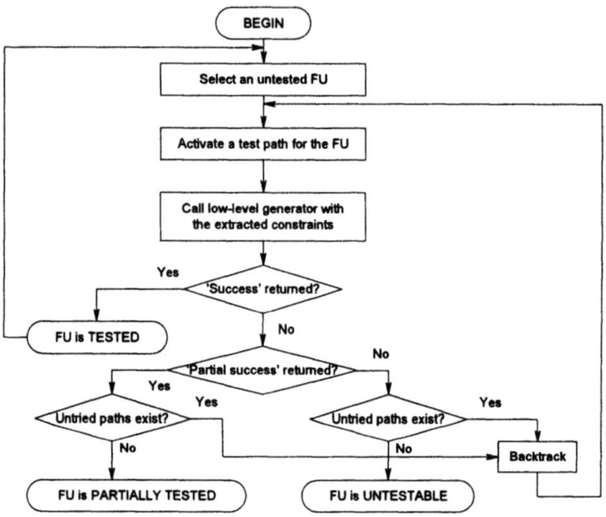

Figure 3. Interaction between High-Level and Low-Level Generators

The datapath test generator has a hierarchical structure. The high-level part of the generator performs symbolic path activation on the RT-level. During the path activation, functional constraints are extracted which will be applied to the low-level test generator. The tasks of the latter are to generate gate-level tests for the functional units (FU) and to assemble the final test of the datapath.

Test generation for the data path takes place in the following way. Tests are created sequentially for each functional unit. Justification and propagation constraints are extracted at the high level and passed to the lower level test generator. During constraints extraction for the target FU, for all non-target FUs, functional information is applied to perform propagation and justification at the functional level. Such an information, in form of simplified behaviour of the block is preliminarily extracted and recorded in a special *transparency library*. This information will consist of a set of input/output mappings (so called I-paths [13] and F-paths [14]).

The low-level test generation process consists of two stages. In the first stage, input values are generated to satisfy the high-level conditional constraints. This task can be treated as a typical constraint satisfaction problem (CSP) [15]. In the second stage, random values are generated and simulated through propagation constraints to derive input patterns for structural level fault simulation of the FU under test. High-level test generator calls the low-level generator repeatedly in a loop. In general case, a single activated path is not enough to reach 100 per cent fault coverage for a FU. The test set for a FU can consist of vectors generated via different activated paths, and therefore, via different calls to the low-level generator. Record has to be kept of the faults detected by previous low-level runs. On each call the low-level generator reads and writes the list of currently covered faults and keeps it in a special file.

Untestable faults for a FU are determined in the current approach by the following method. Previous to starting test generation for a FU, deterministic gate-level test generator finds the list of redundant faults in the FU. If some inputs of the FU are directly tied to constant signals, it will be taken into account while determining the untestable faults. Fault efficiency is considered to be 100 per cent if *number of tested faults = total number of faults - number of untestable faults*. Due to possible high-level constraints, 100% fault efficiency may be unreachable. When a path is activated, the high-level generator calls low-level test generator and passes the extracted conditional and propagation constraints to the latter. If the low-level generator satisfies all the path activation constraints and generates test vectors achieving 100% fault efficiency for the FU, 'success' will be returned to the high-level generator. Current functional unit will be considered to be tested and next untested FU will be chosen by the high-level generator. In the case when the low-level generator can not solve the extracted conditional constraints, or if the achieved fault coverage in current FU remains low or unchanged, high level generator will be informed about it and it will try to activate an alternative path for testing the FU. Figure 3 shows the data flow of interaction between both parts of the generator.

4. EXPERIMENTAL RESULTS

Experiments were carried out on two hard to test RT-level sequential circuits: a 8-bit multiplier Mult 8x8 based on the Robertson's algorithm and a Greatest Common Divison (GCD) circuit. Both of the circuits consist of data path and control parts interconnected through global feedback loops, and they have only one register directly observable. The circuits posed difficult test generation problems caused mainly by data dependent global loops.

The experiments were run on a Sun Ultrasparc 1 computer and are presented in Table 1. The actual quality of tests generated was measured at the low-level by applying gate-level fault simulation to the whole circuit. In test generation experiments, only data path was the target of test generation, the rather high fault coverage reached also for the control path was achieved as a side effect. This fact refers to rather strong fault equivalence or dominance relationships between the faults in datapath and control parts of these circuits. The level of 100% fault coverage was not reached because of either redundancy of the embedded blocks or bad testability of the circuit in general.

Table 1.

Circuit name	GCD	Mult 8x8
Number of gate-level faults	1066	4432
Data path gate-level fault coverage (%)	95,1	95,9
Control path gate-level fault coverage (%)	89,4	92,1
Test generation time (s)	37,1	36,2
Number of generated symbolic test patterns	53	93
Total test length (number of clock cycles)	627	2797

For showing the efficiency of the test generator, we refer also to our experiments on the benchmark family of RISC type processors [7]. The family of benchmarks consists of processors which vary in the instruction set (processors with 4, 8 and 16 instructions) and in the bitwidth (4, 8, 16 and 32-bit processors). The benchmark family was created by describing the high-level behaviour of processors in VHDL and by synthesizing the gate-level implementations with SYNOPSYS. Then AG-models were synthesized both for higher (instruction) level and lower (gate) level designs. As an example, for the 32-bit processor with 21152 faults our hierarchical ATPG needed for generating tests 4.3 sec whereas the gate-level ATPG needed for creating tests with the same quality 2584 sec. The results of the experiments showed the efficiency of both the high-level and hierarchical mixed-level approaches as compared to the gate-level approach. The efficiency rises when the complexity (number of instructions and bitwidth) increases.

5. CONCLUSIONS

A new multi-level ATPG for digital systems is presented. The methods implemented are based on using alternative graphs as a uniform model for representing digital systems at different abstraction levels. The uniformity of the model allows to generalize methods developed earlier for the logical level, to higher functional levels as well and to use common procedures throughout the levels.

REFERENCES

[1] H. Fujiwara and T. Shimono, *On the Acceleration of Test Generation Algorithms*, IEEE Trans. Comput., vol. C-32, pp. 1137-1144, December 1983.

[2] P.N. Anirudhan and P. R. Menon, *Symbolic Test Generation for Hierarchically Modeled Digital Systems*, Proc. IEEE International Test Conf., pp. 461-469, Sept. 1989.

[3] J. Lee and J.H. Patel, *ARTEST: An Architectural Level Test Generator for Data Path Faults and Control Faults*, Proc. IEEE International Test Conf., pp. 729-738, Oct. 1991.

[4] M. Karam, R. Leveugle and G. Saucier, *Hierarchical Test Generation Based on Delayed Propagation*, IEEE Int. Test Conf., pp. 739-747, 1991.

[5] M. Gulbins and B. Straube, *Applying Behavioral Level Test Generation to High-Level Design Validation*, IEEE European Design & Test Conference, Paris, 1996.

[6] H. Krupnova and R. Ubar, *Constraints Analysis in Hierarchical Test Generation for Digital Systems*, Proc. of the Baltic Electronic Conference, Tallinn, Oct. 9-14, 1994.

[7] R. Ubar, *Test Synthesis with Alternative Graphs*, IEEE Design&Test, pp.48-57, Spring 1996.

[8] R. Ubar and M. Brik, *Multi-Level Test Generation and Fault Diagnosis for Finite State Machines*, Lecture Notes in Computer Science No 1150. Dependable Computing – EDCC-2, Springer-Verlag, pp. 264-281, 1996.

[9] R. Ubar, *Test Generation for Digital Circuits Using Alternative Graphs*, Proc. Tallinn Technical University, No.409, Tallinn, Estonia, 1976, pp.75-81 (in Russian).

[10] S.B. Akers, *Binary Decision Diagrams*, IEEE Trans. on Comp., Vol.27, 1978, pp.509-516.

[11] S.M. Thatte and I.A. Abraham, *Test generation for microprocessors*, IEEE Trans. on Computers, Vol.29, 1980, pp.429-441.

[12] P.C. Ward and J.R. Armstrong, *Behavioral fault simulation in VHDL*, ACM/IEEE 27th Design Automation conference, 1990, pp.587-593.

[13] M. S. Abadir and M. A. Breuer, *A Knowledge Based System for Designing Testable VLSI Chips*, IEEE Design & Test, Aug., 1985, pp. 56-68.

[14] S. Freeman, *Test Generation for Data Path Logic: The F-Path Method*, IEEE Journal of Solid State Circuits, vol.23, Apr., 1988, pp. 421-427.

[15] B. Sallay, A. Petri, K. Tilly and A. Pataricza, *High Level Test Pattern Generation for VHDL Circuits*, Proc. of the European Test Workshop, Montpellier, June 1996, pp.202-206.

21

PATH SELECTION BASED ON INCREMENTAL TECHNIQUE

S. Cremoux, N. Azemard and D. Auvergne

LIRMM: Laboratoire d'Informatique, de Robotique et de Microélectronique de Montpellier, UMR 5506 UM2/CNRS, 161 rue ADA , 34392 Montpellier Cedex 5, FRANCE

ABSTRACT

This paper addresses the problem of selecting a set of paths to optimize the performance of a combinational circuit. Comparison between different path enumeration algorithms is presented here considering realistic delay values for the different path elements. Application is given on ISCAS'85 benchmarks where the CPU times of the different investigated algorithms are compared.

1. INTRODUCTION

Identifying the performance bottleneck on a circuit is one of the most difficult task to be realized at the last step of the circuit design flow. To satisfy the imposed delay constraints this implies a full path enumeration with a complete account of a real (post layout like) delay evaluation on the different switching blocks. At this level they are generally specified as the maximal clock period at which the combinational circuit is required to operate, or equivalently as the longest delay admissible on a circuit path. In other terms, optimizing the propagation delay on the critical paths constitutes the necessary condition for the implementation of fast circuits, as well as minimizing the size of gates belonging to non critical paths is the natural alternative of power saving implementation techniques [1,2]. It appears then important to develop fast and accurate path identification techniques [3]. This paper addresses the problem of critical path selection for physical level performance optimization. If the path exploration in a circuit can be obtained easily from timing verifiers [4,5], the resulting topological delays, obtained from graph exploring techniques such as BFS or DFS [6] algorithms, can be pessimistic [1,6]. This is a consequence of not considering the logic behavior. Then, many paths may not be sensitizable i.e. no input vector can be found to activate these paths which are considered as false paths. Static sensitization conditions are obtained by searching for the input vector that activates the corresponding paths and the critical path of the circuit is then defined by the longest sensitizable path of the circuit. However, considering the delay of the circuit elements it appears that a statically unsensitizable path may be dynamically sensitized. The delay of the longest dynamically sensitizable path (referred as a true path) defines the real delay of the circuit. Timing analysis techniques which check for the sensitization of every path are computationally expensive and too slow to be used for large circuits. They suffer from the severe problem of path explosion.

In this paper we address the problem of true path selection using complementary path selection and sensitization approaches [7,8,9]. We implement a new incremental technique [10] to identify statically sensitizable critical paths. We compare this path selection algorithm to previously used enumeration techniques using realistic evaluation of the gate delays [11,12].

2. SEARCH PATH ALGORITHMS

We based this work on circuit path selection onto three graph exploring techniques: the breadth-first-search (BFS), the depth-first-search (DFS) and the incremental algorithms [6,13]. The path enumeration is obtained on a graph representation of the complete circuit, where the nodes represent the gates and the edges the connections between components. These algorithms allow to search circuit paths without considering the logic behavior of the circuit.

The key idea of this work is to consider realistic delay values. These values have been evaluated from post layout circuit extraction using analytical expression of delays formerly developed [11] for submicronic processes. We considered here a 0.7μm transistor effective length CMOS process. Validation of the observed longest path is then obtained using standard static sensitization technique [14].

2 - a BFS algorithm is one of the simplest algorithms for searching a path [13]. It works by successive procedures searching for all the gates at a logical depth from a source node (a primary input). Then it successively processes, for all input nodes, all the gates with increasing depth from this source. At each node the path is constructed, considering all the available directions to the successor nodes. Each incomplete path of the list is then updated by all the considered successors. At the primary output all the paths are built at the same time. The longest paths are obtained by sorting the resulting complete path list.

2 - b DFS algorithm is a recursive algorithm [13] that traces all circuit paths through the graph representation of the circuit. Paths are successively built from primary input to primary output by considering all the successor nodes of each path under process. A path is recorded when no more successor is available. Then the procedure initiates a new path from the last divergence node of the preceding one.

In these enumeration techniques all the circuit paths are considered and stored. This implies prohibitive CPU time and memory resources for large circuits. Speed up of this algorithm has been proposed [1] limiting the path enumeration to the paths that have a delay greater than a threshold value. In fact this path number reduction is efficient for a well defined threshold which is specific of the circuit.

DFS method identifies all the paths, and sorts the circuit critical path at the expense of important CPU time which becomes quickly prohibitive for complex circuits. BFS algorithm appears faster but clearly underestimates the value of the critical path. However no complete path enumeration, allowing speed-power trade-off on the different branches can be performed.

2 - c INCREMENTAL technique has been adapted in order to apply optimization criteria for power and delay on combinational circuit paths: it gives the enumeration in a non increasing order of delays of the longest paths for a given acyclic directed graph.

The search is realized associating to each node and edge a data structure constituted by two parameters:

- the max delay to sink (MDS),

- the branch slack (BS).

The MDS parameter represents the maximum propagation delay time evaluated from the output of a gate to the output of its successors. The BS parameter is the delay difference between two edges of a divergence branch.

For each node the *Max Delay to sink* value is calculated, allowing the determination of the maximum delay between primary inputs and outputs.

Representing all the primary inputs by a source node the maximum delay of the circuit is then obtained from the *Max Delay to Sink* value.

We initiate at 0 value the *Max Delay to Sink* value of the primary outputs, which is then calculated for each primary output node as:

$$MDS_{gate} = Max_{down\text{-}stream\ gates}\ (MDS_i + Max\ (T_{HL}, T_{LHi})\ )$$

where T_{HLi}, T_{LHi} represent the elementary fall and rise delay time, respectively.

Moreover, to obtain a path enumeration in the increasing order, we associate to each edge a value allowing to control the path search. This value, the *Branch Slack* (BS), represents the difference of delay between two edges or equivalently between two possible paths. We calculated the *Max Delay to Sink* value (MDS_{max}) of each edge from the *Max Delay to Sink* and the delay values of each successor (MDS_{calc}). Edges are then sorted in the increasing order of the difference between MDS_{max} and MDS_{calc} values. The *Branch Slack* value of each edge is then obtained from the difference between the MDS_{calc} current edge and the MDS_{calc} preceding edge values (used as the reference).

$$BS = MDS_{ref} - MDS_{edge}$$

The BS value is set to 0 on the edge with the greater MDS value.

These two parameters are associated to all the graph elements as shown in Figure 1.

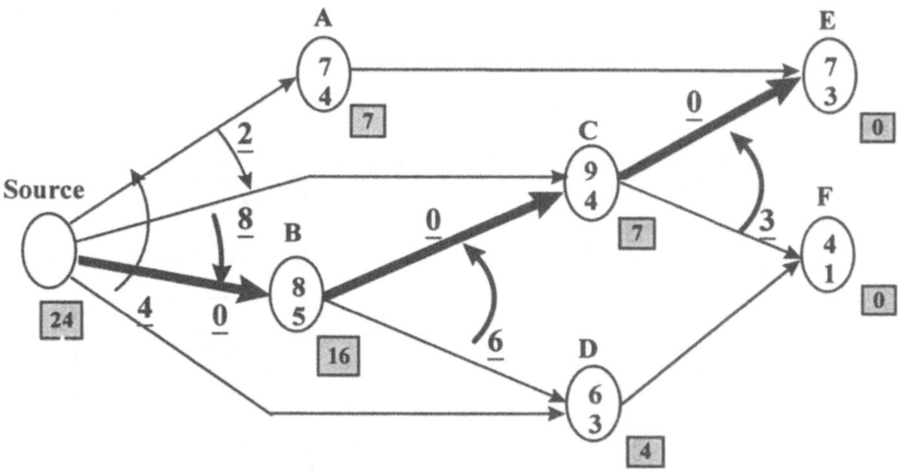

Figure 1. Illustration of the graph representation of the C17 benchmark circuit, propagation delays (Thl,Tlh) are given for each node xx stands for MDS and x for BS values, respectively.

To realize the path enumeration of a definite number of paths we implemented the incremental technique as shown in Figure 2.

```
ListPaths
        Path
        node
        edge
        Path = generate_long_path()
        ListPaths.append (Chemin)
        while (number_path < K)
        Path = ListPath.get_max_next_delay ()
        edge = Path.get_next_successor()
        while (edge ≠ 0)
                node = successor (edge)
                Path.append (node)
                edge = First_Edge (node)
                Path.append_successor (Edge_Succ (edge))
                end_while
                Path.sort_successor()
                Path.calcul_next_delay()
                ListPaths.append()
        end_while
```

Figure 2 : INCREMENTAL technique algorithm implemented on SPARC 20.

Processing the circuit graph representation by considering increasing order of branch slacks, this technique sorts directly the path with the greatest delay (0 branch slack value) and the paths with immediately inferior delay values (by increasing order of branch slacks).

Table 1 summarizes the path enumeration obtained on the circuit C17 given in Figure 1 where the longest path is shown with bold lines (B,C,E path of Table 1).

Table 1. Complete path enumeration for the C17 circuit.

iteration	Path	Delay (ns)	Sorted successors (*Branch Slack*)
1	B, C, E	24	F(3), D(6), C(8)
2	B, C, F	21	
3	B, D, F	18	
3	C, E	16	A(2), F(3)
5	A, E	14	D(4)
6	C, F	13	
7	D, F	10	

3. RESULTS AND CONCLUSION

We applied these three algorithms (BFS, DFS and incremental technique) to ISCAS'85 combinational circuits. The incremental technique has been limited to the 1000 and 10000 longest paths. Results of the comparison of the performances of the different techniques for the longest path classification, in term of CPU time, are given in Table 2.

Table 2. Comparison of BFS, DFS and incremental techniques; the path number and necessary CPU time (in seconds) are given.

			CPU	Time		
					Incremental	Technique
Circuit	gates nb.	Path nb.	DFS	BFS	1 000 paths	10 000 paths
c880	529	8642	0,68	0,7	1,13	11,4
Adder16x2	328	36304	4,05	6,73	2,72	37,2
c432	249	291826	26,8	29,5	1,43	78
c499	700	397888	42,2	130	1,27	52
c1908	1075	729057	insufficient memory	insufficient memory	1,28	35,9
c1355	628	4173216	insufficient memory	insufficient memory	1,5	45,7

As shown and except for small size circuits, the incremental technique appears always faster in searching for the longest paths in large circuits. Due to the important necessary memory allocation, BFS and DFS algorithms become impractical for large circuits.

In conclusion these results show clearly the interest of using incremental technique for circuit path analysis. The control of the number of paths to be investigated gives to this technique a preponderant advantage with respect to the BFS and DFS algorithms. The resulting reduction in required CPU time and memory allocation allow to treat large circuits. Using real delay time in path evaluation and path ordering gives large facilities in circuit verification and optimization. This can be obtained for path classification in increasing or non increasing order of delays as well than for power or in trading speed for power.

An example of application is given in Figure 3 where we plot for the C1908 circuit (1075 Gates and about 730 000 Paths), the number of paths ordered with respect to the delay, for a standard implementation (A curve) with all the transistor widths at 2µm (0.7µm ATMEL ES2 process: Ecpd07) and the new delay profiling (B curve) resulting from the application of optimizing techniques [18] to control the fan out of the gates.

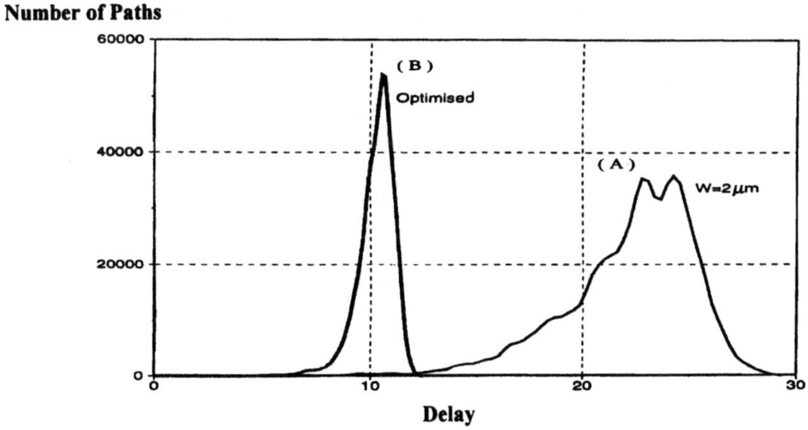

Figure 3. Illustration of the number of paths for the circuit with all W=2µm (curve A) and for the same optimized circuit (curve B).

As shown only optimizing the twenty five paths (among all the 730 000 paths which have been classified) with the greatest delay is sufficient to improve significantly the speed performance of the circuit. This has been obtained for speed improvement by selecting heavily loaded nodes. Selecting paths with the shortest delay for power optimization may be considered in the same way by down sizing the gates of these paths. Power delay optimization, considering both methods is under development.

REFERENCES

[1] D.H.C. Du, S.H.C. Yen and S. Ghanta, *On the General False Path Problem in Timing Analysis*, ACM / IEEE Design Automation Conference, pp. 555-560, June 1989.

[2] H.C. Chen and D. Du, *Critical path selection for performance optimization*, ACM / IEEE Design Automation Conference, Los Alamitos, California, pp. 547-550, 1991.

[3] J. Benkoski, E.V. Meersch, L. Claesen and H. De Man, *Efficient Algorithms for Solving the False Path Problem in Timing Verification*, IEEE International Conference on Computer-Aided Design, pp. 44-47, June 1987.

[4] N.P .Jouppi, *Timing analysis and performance improvements of MOS VLSI designs*, IEEE trans. on CAD, vol CAD 6, n°4, 1987.

[5] J.K. Ousterhout, *CRYSTAL : a timing analyser for NMOS VLSI circuits*, Proc. 3rd Caltech VLSI conf. R.BRYANT, ed. 1983.

[6] P.C. McGeer and R.K. Brayton, *Efficient Algorithms for Computing the Longest Viable Path in a Combinational Network*, Design Automation Conference, pp. 561-567, June 1989.

[7] H.C. Chen and D. Du, *Path Sensitization in Critical Path Problem*, IEEE Transactions on CAD of Integrated Circuits and Systems, vol. 12, n°. 2, pp. 196 - 207, February 1993.

[8] H.R. Lin and T. Hwang, *Dynamical Identification of Critical Paths for Iterative Gate Sizing*, IEEE International Conference on Computer-Aided Design, pp. 481-484, June 1994.

[9] R. Peset Llopis, *Exact Path Sensitization in Timing Analysis*, PATMOS'94, pp. 22-29, September 1994.

[10] Yun-Chen Ju and R.A. Saleh, *Incremental techniques for the identification of statically sensitizable critical paths*, Proc. 28th Design Automation Conf., pp.541-546, 1991.

[11] D. Auvergne, D. Deschacht and M. Robert, *Input waveform slope effects in CMOS delays*, IEEE Solid State circuits, vol. 25, n°6, dec. 1990.

[12] J.M. Daga, S.Turgis and D. Auvergne, *Inverter delay modeling for submicrometre CMOS process*, IEE Electronic Letters, vol.32, n°22, pp.2070-2071, October 96

[13] S. Yen, D. Du and S. Ghanta, *Efficient Algorithms for Extracting the k Most Critical Paths in Timing Analysis*, Design Automation Conference, pp. 649-654, June 1989.

[14] A. Dargelas, C. Gauthron and Y. Bertrand, *MOSAIC: multiple strategy oriented sequential ATPG for integrated circuits*, ED&TC, pp.29-36, Paris 1997.

[15] J.P. Silva, K.A. Sakallah and L.M. Vidigal, *FPD-An Environment for Exact Timing Analysis*, IEEE International Conf. on Computer-Aided Design, pp. 212-215, November 1991.

[16] A. Saldanha, H. Harkness, P.C. McGeer, R.K. Brayton and A.L. Sangiovanni-Vincentelli, *Performance Optimization using Exact Sensitization*, ACM / IEEE Design Automation Conference, pp. 425-429, June 1994.

[17] S. Devadas, K. Keutzer and S. Malik, *Delay Computation in Combinational Logic circuits: Theory and Algorithms*, ACM / IEEE International Conference on Computer Aid design, Santa Clara, California, pp. 176-179, 1991.

[18] S. Turgis, N. Azemard and D. Auvergne, *Design and selection of buffers for minimum power-delay product*, ED&TC, pp.224-228, Paris, March 1996

22

CHIP AREA ESTIMATION FOR SC FIR FILTER STRUCTURES IN CMOS TECHNOLOGY

Adam Dąbrowski and Rafał Długosz

Division for Signal Processing and Electronic Systems
Poznań University of Technology
ul. Piotrowo 3a, 60-965 Poznań
POLAND

ABSTRACT

This paper presents evaluation and comparison of different SC FIR filter structures with the point of view of a chip area for 3 μm, 2 μm, 0.8 μm CMOS technologies. To realize the considered structures, three types of building blocks are necessary, namely: memory elements (which store a signal sample over one or more clock periods), multipliers (which realize multiplication of signal samples by a constant coefficient), and summers (which add two or more processed signal samples). Each of these basic building blocks comprieses elements of only three types: operating amplifiers (OA's), switches (S's) and unit or coefficient capacitors (UC's or CC's, respectively). To estimate the entire chip area for the considered FIR structures, it is necessary to determine the numbers of particular elements as functions of the filter order N.

1. INTRODUCTION

In seventies a revolution began in the integration of electronic filters with achievements in MOS VLSI technology. It became to be possible to integrate perfect switches, precise capacitors (with respect to capacitance ratios), and satisfactory op amps (OA's). This led to entirely new class of analog circuits - the switched-capacitor (SC) circuits [1]. Realization of filters was possible e.g. by replacement of resistors by configurations of switches and capacitors. The SC technique possesses many important advantages. The most important of them are: low power consumption for applications with low dynamic range, simple structure (low production costs) and large degree of parallelism in the realization of certain signal processing tasks. Thus, SC circuits can be preferable even for processing of signals at relatively high frequencies (in the range of MHz for CMOS technology). In SC technique, the filter length is, in general, independent on the maximum frequency but it is limited by the maximum chip area and by the minimum signal-to-noise ratio (SNR) [1,2]. The chip area depends on the technology e.g. 0.8μm 2 μm or 3μm CMOS. There exist four basic FIR filter structures such as: tapped delay line structure, reversed delay line structure, parallel structure and rotator structure [1]. Each of these basic structures possesses specific advantages as well as disadvantages. Therefore it is often necessary to search for a compromise between different possibilities. Polyphase decomposition can be a proper solution to this problem. It offers different combinations of some basic structures [1,2,5,6,8]. In this paper, the basic structures, as well as, structures based on the idea of polyphase decomposition are considered and they are compared using estimation of a chip area as the evaluation criterion.

In next Section we present some important properties of the basic structures, i.e., their advantages as well as disadvantages. In third Section, the method for evaluating the chip area of SC FIR filter structures is described. Finally a short summary is given.

2. BASIC FILTER STRUCTURES

We consider SC FIR filter structures presented in [1]. These are:

- **Delay line structures:** They can be realized either as a tapped delay line structure or as a reversed delay line structure. Both of them can be realized parallelly (operations "data read" and "data write" are performed in the same phases in all delay elements) or serially (i.e., one after the other, thus composing a single filter operation cycle). In the first case, a large number of OA's, and a small number of clock phases is necessary. Such structures can be based either on an even-odd delay element [1,4] (structure No. 1) or on Gillingham delay element (structures No. 2 and 3). Advantages and disadvantages of each of these structures follow, in particular, from properties of the even-odd and the Gillingham delay elements. Among advantages of the even-odd delay element are: compensation of the offset-voltage and insensitivity to the capacitor mismatch. The last property follows from the fact that U_{out} is independent from the values of capacitors. Structure No. 1 is driven by a four phase clock but with only two clock phases per sample. On the other hand, the Gillingham element requires only a two phase clock. The disadvantage of structures No. 2 and No. 3 is their sensitivity to capacitor mismatch and to offset-voltage. Nevertheless, the offset errors do not accumulate along the delay line. In the case of serial realizations, only a single OA is sufficient but a complex multiphase clock is required (delay line structures No. 4 and 5). Structure No. 5 is a reversed delay line. The main disadvantage of all delay line structures is their sensitivity to overwrite errors accumulated along the line during the "data read" and "data write" operations.
- **Rotator FIR structure:** This structure comprises the so-called rotator switch that connects signal samples with capacitors of the summer circuit. The signal samples are stored in sample-and-hold elements. Structures of this kind are characterized by large numbers of OA's and many switching phases.
- **Parallel FIR structure:** In this structure, signal samples are stored in individual capacitors. The advantages are: no cumulative errors (in opposition to the delay line structures) and a single active element only. The disadvantages are: a very large number of switching phases: $2(N+1)$, where N is the filter order, and a very large number of capacitors, i.e., about $(N+1)^2$.

3. CHIP AREA ESTIMATION

To estimate the required chip area for a particular FIR structure, it is necessary to determine the numbers of elements as a function of the filter order N. Then we consider the following equation

$$S(N) = S_{\mathrm{OA}} X_{\mathrm{OA}}(N) + S_{\mathrm{UC}} X_{\mathrm{UC}}(N) + S_{\mathrm{CC}} X_{\mathrm{CC}}(N) + S_{\mathrm{S}} X_{\mathrm{S}}(N) + S_{\mathrm{C}}(N) P \quad (1)$$

where:

$S(N)$ - the entire chip area,
N - filter order,
S_C - area of a single connection,
P - number of clock phases,
S_{OA} - area of a single OA,
$X_{OA}(N)$ - number of OA's,
S_{UC} - area of a single unit capacitor (UC),
$X_{UC}(N)$ - number of UC's,
S_{CC} - area of a single coefficient capacitor (CC),
$X_{CC}(N)$ - number of CC's,
S_S - area of a single switch (S),
$X_S(N)$ - number of S's

Assuming that the chip shape is approximately square (see Figure 1), we can write

$$S_C = k\sqrt{S_{OA}X_{OA} + S_{UC}X_{UC} + S_{CC}X_{CC} + S_S X_S} = kb \tag{2}$$

Coefficient S_C results from the necessity of assignment of some part of the chip area for connections. This area depends linearly on the number of clock phases P. Coefficient k depends on the technology used, b is the length of the chip equal to the square root of the area of all filter elements. Coefficients S_{OA}, S_{UC}, S_{CC}, S_S depend also on the technology. We use their estimated values. Values for parameters X_{OA}, X_{UC}, X_{CC}, X_S are computed individually for all considered structures. To derive and evaluate all realizable and reasonable compositions of the basic SC FIR filter structures, we follow the so-called morphological approach [3].

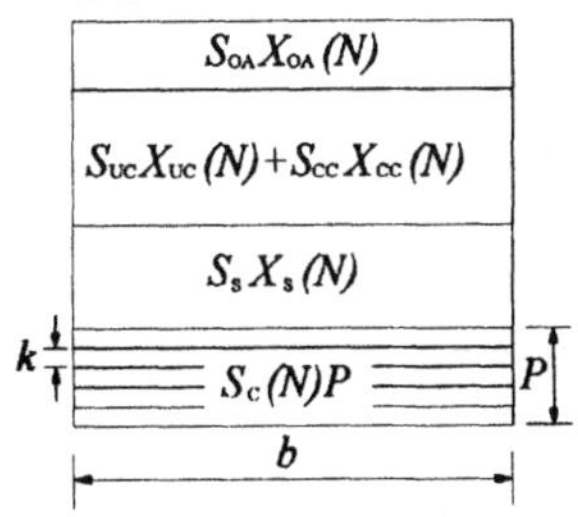

Figure 1. Organization of the chip area

4. DERIVATION OF DIFFERENT FILTER STRUCTURES

In this Section we introduce the so–called "morphological approach" [3] to the derivation of all realizable (but reasonable) composite FIR filter structures, i.e. those which are composed of at least two basic structures. Thus, together with all basic structures, we get a reach collection of different SC FIR filter structures. The morphological approach consists of the five following steps:

1.Problem formulation: We are going to realize composite SC FIR filter configurations using the basic SC FIR filter structures introduced above. Then we shall qualify them according to the following criteria: number of circuit elements, chip area, operation speed, network performance, clock complexity, etc.

2.Characterization of fundamental elements: The following four basic SC FIR structures serve as the fundamental elements for the so-called "morphological box": multi–op–amp delay line FIR structure, multiphase delay line FIR structure, parallel (multi–C) FIR structure, rotator FIR structure.

3.Derivation of a morphological box: In our case this is a 4–dimensional box with four basic structures (those listed above) in each dimension. The possible solutions correspond to different positions in this box. Thus, theoretically, we could get composite structures composed of even four basic structures but we will stop on two or at most three of them.

4.Evaluation of solutions contained in the morphological box: Many of the theoretically possible solutions are not realizable or not reasonable in practice. Thus we eliminate them using the following rules:

- In order to reduce the number of op amps, the multi–C FIR structure should be applied as the last link of a composite filter structure or followed by a delay line with recharge branches and recharge summer circuits. Signal samples (charges on capacitors) can be read only once.
- A particular basic structure can (in practice) occur only once in a composite structure. Two consecutive identical basic structures can be combined into one. Configurations composed of more than two basic structures are of little practical concern.
- Structures equivalent to a tapped delay line are suitable for the first link, while structures realizing the reversed delay line are well fitted for the last link.

Taking the above conditions into account, we obtain 8 reasonable composite FIR SC configurations, which are compared with the basic structures in Figures 2, 3 and 4.

5.Evaluation of the solutions: Selection of superior solutions among the FIR SC structures obtained in step 4 can be based on the evaluation of their feasibility as integrated circuits and on their performance in applications.

5. CONCLUSIONS

The chip area is presented in Figures 2, 3 and 4 as a function of the filter order N for 3μm, 2μm and 0.8μm CMOS technologies, respectively, and for FIR filter structures described in previous Sections.

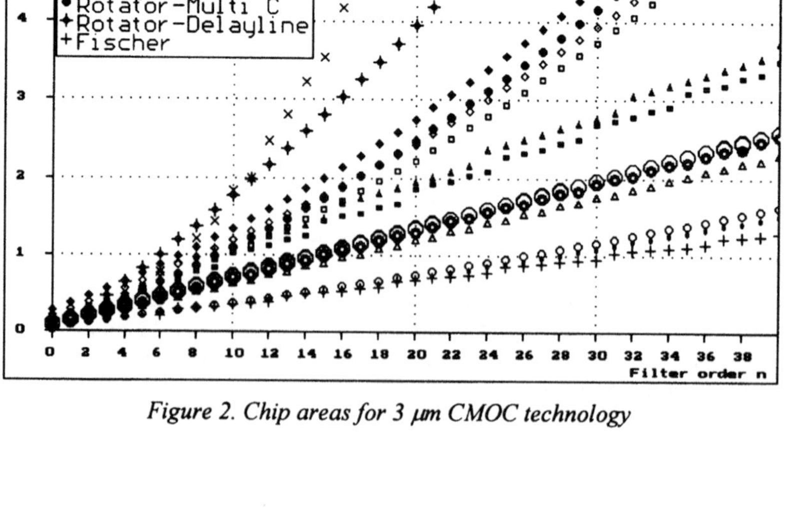

Figure 2. Chip areas for 3 μm CMOC technology

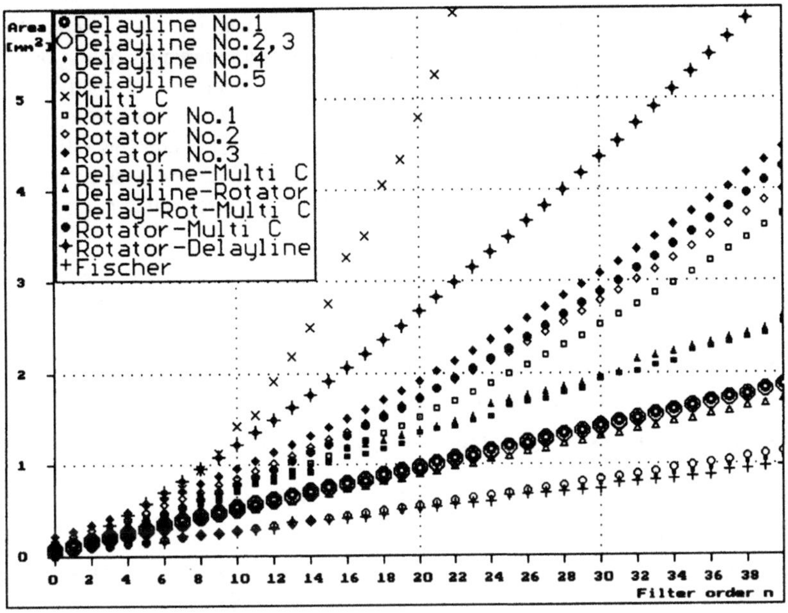

Figure 3: Chip areas for 2 μm CMOC technology

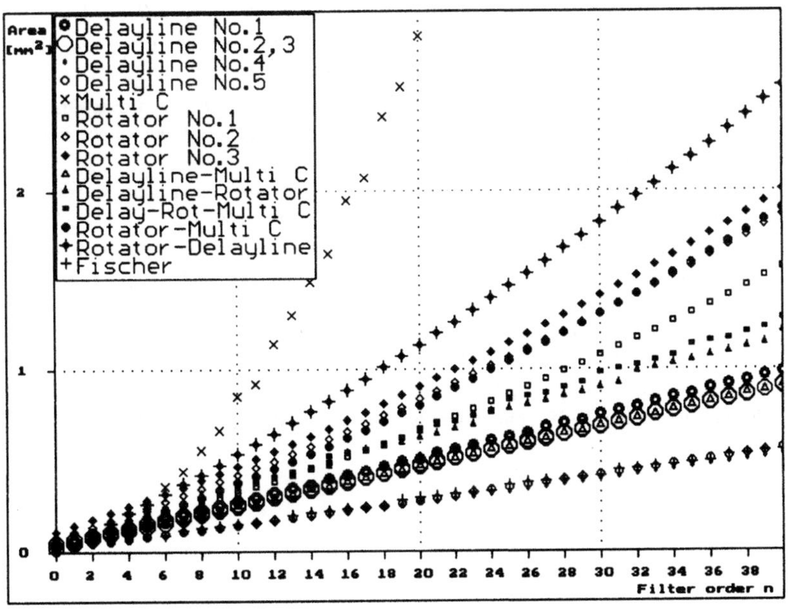

Figure 4: Chip areas for 0.8 μm CMOC technology

If we improve the technology, the area of particular elements will decrease but not linearly with the technology parameters. In consequence, the entire chip area will decrease. The greatest area change can be noticed for OA's (3.7:1 for the change from 3μm to 0.8μm CMOS technology). Intermediate change can be observed for switches (3.33:1). The change of the area of capacitors is smaller, i.e., about 1.8 : 1. In result the highest progress appears in these structures which comprise larger numbers of OA's (rotator, delay lines No. 1, No. 2, No. 3) and smaller numbers of capacitors. Thus some structures preferable,e.g., from the chip area point of view in one technology may be not optimum in the other one. For example, the chip area of a multi-C structure (N = 20) is equal to 6.2mm^2 for 3μm CMOS technology but 2.9mm^2 for 0.8μm CMOS technology. It means that the chip area can is decreased by 2.1:1. For the rotator structure this ratio is greater. It equals 3.2:1. Rotator structure comprises a large number of OA's and, therefore, the improvement is bigger than in the case of the multi-C structure, which comprises a large number of capacitors but only a single OA.

A lot of freedom exists in the design of composite structures. A filter of order N can be realized in many different ways, e.g., the order of the rotator-delay line structure is given by $N = M + L$. One of values M and L can be treated as a parameter. The chip area in every case will be different. To evaluate the optimal chip area, function $S = \mathrm{f}(M, L)$ must be calculated to find the best partition of N to M and L. This function depends on technology parameters S_{OA}, S_{UC}, S_{CC}, S_S. In result, the optimum partition of N changes with the technology.

REFERENCES

[1] A. Dąbrowski, *Multirate and multiphase switched-capacitors circuits*, Chapman & Hall, London, New York 1997

[2] A. Dąbrowski, U. Menzi and G.S. Moschytz, *Design of switched-capacitors FIR filters to a low-power MFSK receiver*, IEE Proceedings - G, Vol. 139, No. 4, 1992, pp. 450-466

[3] G.S. Moschyz, *The morphological approach to network and circuit design*, IEEE Trans. Circuits Syst., Vol. CAS -23, No. 4, 1976, pp.239-242

[4] A. Dąbrowski, U. Menzi and G.S. Moschytz, *Offset-compensated switched-capacitor delay circuit that is insensitive to stray capacitance and to capacitor mismatch*, Electron. Lett, Vol. 25, No. 6, pp. 387-389

[5] Z. Ciota, A. Napieralski and J.L. Noullet, *Analog interpolated finite impulse response filters*, Proc. European Conf. Circuit Theory Design, Davos, Switzerland, 1993, pp. 1367-1372

[6] Z. Ciota, *Theory and practical realization of analog integrated filters particularly taking into account finite impulse response filters*, Łódź University of Technology Press, Łódź 1996 (in Polish)

[7] G. Fischer, *Analog FIR filters by switched-capacitor techniques*, IEEE Trans. Circuits Syst., Vol. CAS-37, No. 6, 1990, pp. 808-814

[8] J.E. Franca and S. Santos, *FIR switched-capacitor decimators with active-delayed block polyphase structures*, IEEE Trans. Circuits Syst., Vol. CAS-35, No. 8, 1988, pp. 1033-1037

23

SIMPLIFIED MODELS OF IC'S FOR THE ACCELERATION OF CIRCUIT DESIGN

Vladimir A. Koval, Mykola B. Blyzniuk and Irena Y. Kazymyra

CAD Department
State University "Lviv Polytechnic"
12 Bandera Str., 290646 Lviv
UKRAINE

ABSTRACT

We consider the problem of time expenses for IC's parameter optimization. The use of the efficient simplified models is motivated. Simulation technique using models of two types of complexity, i.e. exact and simplified models, is proposed. Simplified models are formed automatically on the basis of the theory of design of computer simulation experiments. We present practical example of the simplified model development for IC's parameter optimization.

1. INTRODUCTION

Iterative procedures of circuit design are known to be computation-intensive and the amount of computation grows with the increase of circuit dimensions. Hence the total time of computation (Tm) is the main limitation when rising the complexity of the designed IC's [1,2]. The experience of optimization problems solution shows that if time of objective function calculation is measured in minutes (Tm >1 min) then the parameter optimization procedure will last many hours. So in most cases we shall have to be contented with quasioptimal solution.

In circuit design time of objective function calculation is equivalent to the execution time of circuit analysis program. The exact mathematical models (we call them the 2nd type models) are used in circuit analysis programs. In many cases, e.g. early stages of design or evaluation of compromise solutions, qualitative character of solution is more important for comparative evaluation than exact solution value. For these cases the technique of mixed use of the exact models, i.e. models of the 2nd type, and the efficient simplified models, i.e. models of the 1st type, are proposed by the authors. Models of the 1st type are designed on the basis of the 2nd type models with the use of the theory of design of computer simulation experiments [3,4].

2. PROBLEM OF TIME EXPENSES FOR ICS PARAMETRIC OPTIMIZATION: ONE OF THE VIEWPOINTS

On the whole the problem of IC's multi-objective parameter optimization comes to the problem of constrained single-objective optimization with additive criterion function [1]. This optimization problem is formulated as follows:

Find: $$x^* = \{x_1^*, x_2^*, \dots, x_N^*\}^T \quad (1)$$

minimizing $$\underset{x \in R_{opt}}{Q(x)} = \sum_{k=1}^{M} W_k Q_k(x) \quad (2)$$

such that: $$H(x)=0,\ G(x)\geq 0 \quad (3)$$

where:

$x=\{x_1,x_2,\dots,x_N\}$ - N-dimensional vector of optimization variables (parameters of circuit elements);
x^* - vector of optimal values of parameters;
$Q(x)$ - additive objective function;
$Q_k(x)$ - k-th optimality criterion;
W_k - weight coefficient of the k-th criterion;
R_{opt} - region of acceptability formed by the equality and inequality constraints ($H_j(x)=h_j(x)=0$, $j=1,2,\dots,J$; $G_l(x)=g_l(x)\geq 0$, $l=1,2,\dots,L$).

The formulas (1-3) are considered as the optimization model.

Here arises the necessity of preliminary evaluation of time expenses for solving multiple criterion optimization of IC's parameters. There is no unique answer for the question how to perform such estimation [5]. One of the possible ways of preliminary evaluation is suggested below.

To evaluate the efficiency of the parameter optimization procedures in circuit design we propose the use of special test function, i.e.

$$Q(x) = \sum_{j=1}^{J}(f_{given\,j} - f(x)_j)^2 = \sum_{j=1}^{J}\{f_{given\,j} - \sum_{i=2,4,..N}[a_i(b_i x_{i-1} - c_i x_i^2)^2 + (d_i - p_i x_i)^2]\}^2 \quad (4)$$

The specific features of the developed function are following: a) it has changeable number of parameters xi, this allows to evaluate the effectiveness of the optimization procedure depending on vector x dimensions, i.e. number of parameters of circuit elements; b) it is traditional criterion function providing maximal approximation of the output characteristic of a circuit to the given characteristic; c) this is nonlinear function with ravine-like character.

The effectiveness / problem dimensions dependencies for Rosenbrock and Nelder-Mead optimization methods are presented in Figure 1. This figure shows that in case when time for Q(x) calculation is greater than 1 min the procedure of parameter optimization of a circuit with 20 parameters will take approximately 200 hours.

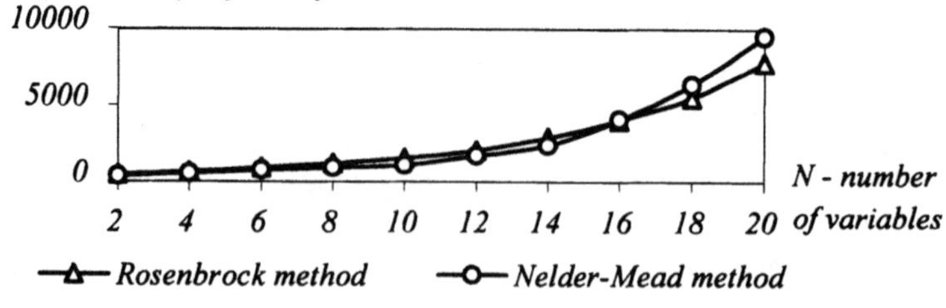

Figure 1. Objective function calculations / number of variables dependencies

When using additive objective function to combine multiple objectives into a single one, optimality of compromise solution of the problem will depend on the weight coefficients values (W_k, $k=1,2,...,M$). Here arises the necessity of comparative evaluation of compromise solutions with different weight coefficient values. The computer time necessary to solve this problem with the use of the exact models, i.e. the 2nd type models, rises considerably.

3. DEVELOPMENT OF THE EFFICIENT SIMPLIFIED MODELS

To increase the efficiency of formulated problem solution we propose to use the simplified models, namely to approximate the additive objective function $Q(x)$ in the region of acceptability by polynomial model with the use of orthogonal polynomials:

$$Q(x) = b_0 + \sum_{i=1}^{k} b_i x_i + \sum_{i=1}^{k} b_{ii} x_i^2 + \sum_{i \neq j} b_{ij} x_i x_j + \sum_{i \neq j \neq g} b_{ijg} x_i x_j x_g + .. + b_{ijg...k} x_i x_j x_g ... x_k \quad (5)$$

The approximation is accomplished by the use of the theory of design of computer simulation experiments [3, 4]. That is to say full or fractional factorial experiment is carried out. The experiment is performed with the model of the 2nd type realized as the circuit analysis program.

Automatic process of forming the simplified models (models of the 1st type) is presented in Figure 2. The following problems of experiments design are being solved automatically:

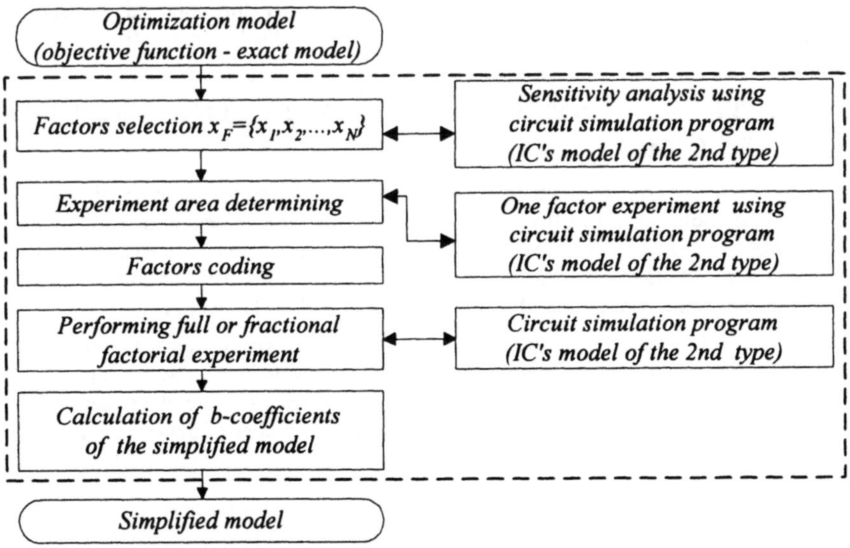

Figure 2. Diagram of forming the simplified models automatically

A. **Selection of the factors from the set of IC's elements parameters (x vector) which is based on the sensitivity analysis with the use of the 2nd type model.** Using the procedure of sensitivity analysis we calculate the matrix of sensitivity $S_x^Q = \left[S_{x_i}^{Q_k} \right]$, where $S_{x_i}^{Q_k}$ ($k=1,2,...,M$, $j=1,2,...,N$) - coefficient of logarithmic sensitivity of Q_k criterion to the change of x_j parameter of a circuit element. Formation of factors vector $x_F = \{x_1,x_2,...,x_{Nf}\}$ is performed by choosing the parameters which have the greatest influence on the circuit performance.

B. Experiment area defining and factor coding. Technique of one factor experiment is used to solve these tasks. The aim of the experiments is to define the intervals of changing the factors so that the region of acceptability would be covered by the experiment area. Depending on the size of the region of acceptability the principle of factor area scanning can be used [6].

C. Performing full or fractional factorial experiment with factors at two or three levels; computation of polynomial model coefficients. Depending on the accuracy of approximation of response surface full factorial experiment with factors at two levels or central composite design experiment with factors at three levels is carried out. To decrease the amount of calculations and to increase the number of factors fractional factorial experiment is performed. Negligible interactions are defined on the basis of full factorial experiment.

The proposed technique is used for nominal design of IC's and for statistical design of IC's as well. Execution of the computer simulation experiment requires a great amount of calculations, i.e. many-fold analysis of the circuit model. For example, performing the full factorial experiment with factors at two levels for approximation of the function of 10 variables requires 1024 calculations of the circuit model. The efficiency of such approach is provided by the fact that almost all calculations, connected with the computer simulation experiment performing, are made **not more than once**. Only *b*-coefficients of the model are calculated for concrete values of weight coefficients W_k. The optimization of approximated additive function $Q(x)$ is not difficult. Since the obtained model of the 1st type is analytical, it allows to efficiently evaluate optimal solution at different values of weight coefficients.

4. PRACTICAL EXAMPLE OF THE EFFICIENT SIMPLIFIED MODEL OF IC

Generator of the pulses of given form will be considered as an example of the use of proposed technique of simplified models development for acceleration of parametric optimization procedure in nominal design. Nominal design problem consists of assigning values to a set of design parameters not subject to statistical fluctuations so that the circuit performance is optimized, provided that certain specifications are met [1].

Initial version of electrical circuit of pulse generator is shown in Figure 3. Input data for nominal design of the above-mentioned circuit are presented in Table 1. The results of circuit simulation of pulse generator are shown in Figure 4. The developed additive objective function for providing the required output parameters of generator circuit is as follows:

$$Q(x) = \sum_{k=1}^{3} W_k Q_k(x) = W_1(T_{puls_{giv}} - T_{puls}(x))^2 + W_2(T_{del_{giv}} - T_{del}(x))^2 + W_3(V_{a_{giv}} - V_a(x))^2 \quad (6)$$

Time expenses for one time calculation of additive objective function with the use of exact models are equal $T_{Q(x)} = 95$ sec. Time expenses for search of optimal solution using exact models (number of variables is 6) are equal $T_m \approx 25$ hours.

Table 1. Input data for circuit design of generator of pulses of the given form

	Name	Required value	Initial value
Output pulse duration [μsec]	T_{puls}	1.5	0.92
Output pulse delay time [μsec]	T_{del}	0.3	0.28
Output pulsed value [V]	V_a	5	5.2

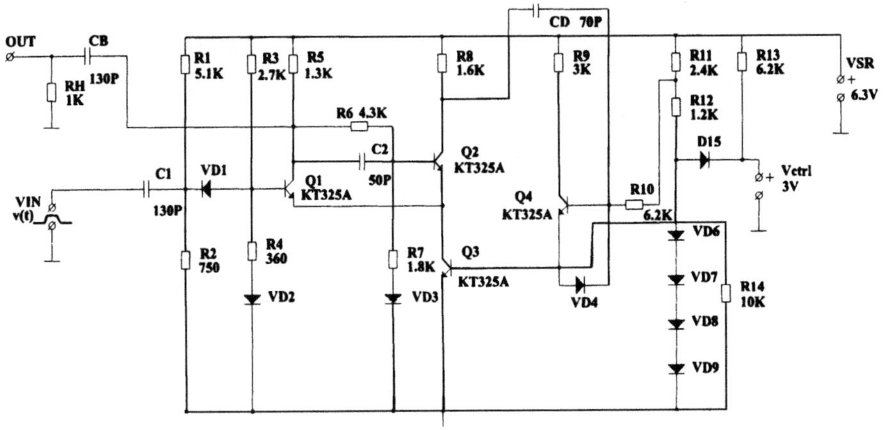

Figure 3. Schematic diagram of the generator of pulses of given form

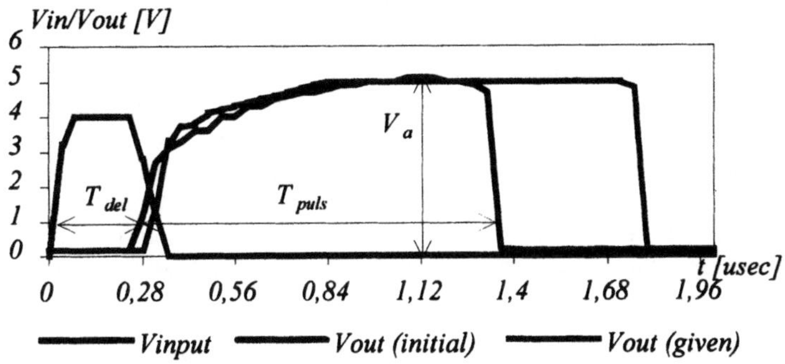

Figure 4. Results of circuit simulation of generator in time domain

According to the proposed technique of simplified models development selection of factors is performed on the basis of sensitivity analysis with the use of circuit simulation program MicroPC [7]. Results of factor selection are presented in Table 2. Experiment area is determined under condition that circuit generates output pulse. The upper and lower levels of factors are calculated automatically. They are ±50% of nominal value.

Table 2. Results of factor selection

Name of the selected element	Nominal value	Sensitivity coefficients S_x^F of the parameters		
		$S_x^{T_{puls}}$	$S_x^{T_{del}}$	$S_x^{A_{puls}}$
R2	750Ω	-0.0188	0.105	0.0008
R5	1.3kΩ	0.0	0.0131	-0.215
R10	6.2kΩ	0.875	0.0	0.089
R11	2.4kΩ	0.412	0.0	0.0424
R12	1.2kΩ	-0.271	0.0	-0.0276
CD	70pF	1.027	0.035	0.094

On the basis of full factorial experiment with factors at two levels we have developed simplified model. This model after normalization and without taking into consideration negligible interactions is following:

$$
\begin{aligned}
Q(\mathbf{x}) = {} & W_1(164.533-(100.0-1.26r_2+0.042r_5+41.98r_{10}+17.279r_{11}-9.78r_{12}+49.31c_D- \\
& -0.498r_2r_{10}-0.155r_2r_{11}+4.37r_{10}r_{11}-3.58r_{10}r_{12}+2.17r_{11}r_{12}-0.123r_2r_{11}r_{12}+ \\
& +2.75r_{10}r_{11}r_{12}-0.13r_2r_{10}r_{11}r_{12}-0.568r_2c_D+19.427r_{10}c_D-0.178r_2r_{10}c_D+6.75r_{11}c_D- \\
& -2.89r_{12}c_D+0.315r_{10}r_{12}c_D+3.28r_{11}r_{12}c_D-0.142r_2r_{11}r_{12}c_D+3.57r_{10}r_{11}r_{12}c_D- \\
& -0.142r_2r_{10}r_{11}r_{12}c_D))^2+W_2(105.487-(100+5.94r_2+0.65r_5))^2+W_3(104.2687- \\
& -(100.0-12.6r_5+5.95r_{10}-1.51r_{11}++7.2r_{12}+1.42r_5r_{10}+0.6r_5r_{11}-1.05r_{10}r_{11}- \\
& -0.37r_5r_{12}+0.28r_{10}r_{12}+0.35r_{11}r_{12}++1.85r_5c_D-2.93r_{10}c_D-0.24r_{10}r_{11}c_D+ \\
& +0.686r_{12}c_D-1.13r_{11}c_D-0.37r_5r_{10}r_{11}c_D+0.302r_{10}r_{12}c_D-0.216r_5r_{10}r_{12}c_D))^2 \qquad (7)
\end{aligned}
$$

Circuit parameter optimization using simplified model with weight coefficients $W_1=W_2=W_3=0.3333$ resulted in the next optimal solution:

$$\mathbf{x}^0=\{R_2,R_5,R_{10},R_{11},R_{12},C_D\}=\{750\Omega,1.3k\Omega,6.2k\Omega,2.4k\Omega,1.2k\Omega,70pF\};\ Q(\mathbf{x}^0)=1402$$

$$\mathbf{x}^*=\{R_2,R_5,R_{10},R_{11},R_{12},C_D\}=\{1.1k\Omega,1.43k\Omega,8.8k\Omega,1.9k\Omega,1.3k\Omega,94pF\};\ Q(\mathbf{x}^*)=10^{-10}$$

Optimal values of optimal output parameters of generator: $T^*_{puls}=1.52\ \mu sec$; $T^*_{del}=0.3\ \mu sec$; $V^*_A=5.043\ V$. Relative error of the obtained optimal solution is $\delta Q=2.8\%$ as compared to the exact solution. Total time saving is $\Delta T_m \approx 20$ hours.

We present the computer time / number of variables dependencies (see Figure 5) for parameter optimization of generator with the use of exact and simplified models. Time for optimization with the use of simplified models includes also the time for their development (using full factorial experiment). To increase time saving, when the factors quantity is greater than 10, fractional factorial experiment should be used. Experiment practice shows that for analog ICs interaction effects of the fifth and higher order are negligible. This allows to carry out fractional factorial experiment by introducing additional factors with a much reduced sample size (as compared to the full factorial experiment).

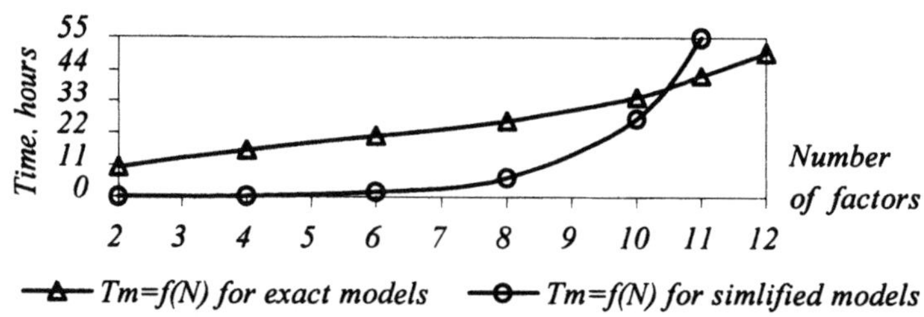

Figure 5. Time / number of variables dependencies

5. CONCLUSIONS

Optimal solution obtained with the use of simplified models can differ from the exact solution. But when performing the analysis of different initial variants of a circuit with the aim of their comparative evaluation, the qualitative character of circuit performance is more important than its precise value. The accuracy of the obtained approximate solution can be improved using the IC's exact model. Time expenses for simplified models development are compensated by significant time saving in IC's parameter optimization procedure, especially when the weigh coefficients of objective function are changeable.

REFERENCES

[1] R.K. Brayton, G.D. Hachtel, A.L. Sangiovanni-Vincentelli, *A Survey of Optimization Techniques for Integrated-Circuit Design,* Proc. of the IEEE, Vol.69, No.10, October 1981, pp.1334-1362.

[2] V.A. Koval, M.B. Blyzniuk, I.Y. Kazymyra, *Acceleration of Circuit Design: One of the Approaches,* Proc. of the 3rd Advanced Training Course on Mixed Design of Integrated Circuits and Systems, Lodz, Poland, 30 May - 1 June, 1996, pp.113-118.

[3] R. Jain, *The Art of Computer Systems Performance Analysis,* Wiley, 1991.

[4] T.H. Naylor (ed.), *The Design of Computer Simulation Experiments*, Duke University Press, Durham, N.C., 1979.

[5] D.M. Himmelblau, *Applied Nonlinear Programming,* McGraw-Hill, 1982.

[6] R.E. Shannon, *System Simulation. The Art and Science,* Prentice-Hall, N.J.,1975.

[7] I.Y. Kazymyra, M.B. Blyzniuk, M.V. Lobur, *Circuit Simulation Program for Training Specialists in the Field of Circuit Design Software Development*, Proc. of the 4th International Workshop on Mixed Design of Integrated Circuits and Systems, Poznan, Poland, 12-14 June, 1997, pp.673-678.

24

LOW POWER METHODOLOGIES FOR GaAs ASYNCHRONOUS SYSTEMS

[1]Stefan W. Lachowicz, [1]Kamran Eshraghian, [2]Jose F. López, [2]Roberto Sarmiento and [3]Hans Jörg Pfleiderer

[1]Edith Cowan University
Joondalup WA 6027
AUSTRALIA

[2]University of Las Palmas de Gran Canaria
35017-Las Palmas de Gran Canaria
SPAIN

[3]University of Ulm
D-89069 Ulm
GERMANY

ABSTRACT

Event Driven Logic is a means of control path implementation for asynchronous digital circuits. This paper presents efficient Gallium Arsenide implementations of the fundamental EDL gates using newly introduced, GaAs pseudo-dynamic latched logic family (PDLL) primitives. The circuits are characterised by very high speed and lower power dissipation compared to their DCFL counterparts.

1. INTRODUCTION

Clock skew is the speed limiting factor in digital synchronous systems. Moreover, the clock-distribution system can dissipate a considerable amount of power, reaching up to 40% of the total power dissipation of the system [1]. On the other hand, self-timed digital systems do not suffer from the problem of the clock skew. However, the penalty is the increased complexity of the system and the requirement to incorporate a handshaking circuitry that permits reliable communications between asynchronous modules. Each module generates an event (in the form of a signal transition) when it is ready to accept data, and another event on completion of its computation. The use of transition signalling is common in self-timed applications due to the time and power savings it allows [2]. The handshaking modules can be implemented using Event Driven Logic described in [3]. Several standard circuit elements commonly required to process transition signals have been developed. These include: Muller-C elements (AND for events), Exclusive-OR gates (EDLXOR), Inclusive-OR gates (EDLINCOR), and inverters. The implementations of theses circuits in CMOS technology are readily available. Especially important is the Muller-C element which is the core building block used in the two- and four-phase handshake protocols [4]. This paper presents an efficient implementation of Event Driven Logic gates in Gallium Arsenide using newly introduced, GaAs pseudo-dynamic latched logic family (PDLL) primitives [5]. PDLL offers many advantages over traditional static GaAs logic families. It allows complex gate design with less power dissipation and, furthermore, it overcomes problems associated with charge degradation in the storage nodes in dynamic logic gates. In addition, it is fully compatible with direct coupled FET logic (DCFL). In Section 2, the principles of Event Driven Logic are summarised. This is followed by the Gallium Arsenide implementations presented in Section 3, and conclusions in Section 4.

2. EVENT DRIVEN LOGIC

Event driven or transition-based logic is an efficient way of approaching the representation and design of asynchronous sequential logic. A detailed description of EDL principles can be found in [3]. Only a brief overview of the basic rules and EDL operators is given here. In EDL concept, the initial conditions of a system are expressed in terms of the logic level assumed by each variable and the subsequent system behaviour is described in terms of the transitions of those variables. The transition is called an event and the lack of transition is called a non-event. If a single line is carrying a logic signal A, there are four possibilities:

(i) ΔA denoting a change in A from 0 to 1
(ii) ∇A denoting a change in A from 1 to 0
(iii) $\overline{\Delta}A$ denoting no change in A at logic 0
(iv) $\overline{\nabla}A$ denoting no change in A at logic 1

Possibilities (i) and (ii) are defined as events, and an event for the signal A can be defined as:

$$\partial A = \Delta A + \nabla A \qquad (1)$$

Possibilities (iii) and (iv) are defined as non-events, and a non-event for the signal A can be defined as:

$$\overline{\partial}A = \overline{\Delta}A + \overline{\nabla}A \qquad (2)$$

There are simple bridging rules between EDL and conventional logic. They are summarised in Table 1.

Table 1. Bridging rules between EDL and conventional logic

EDL	Conventional	Description
$\nabla A + \overline{\Delta}A$	$\overline{A}$	A becomes 0 or remains at 0
$\Delta A + \overline{\nabla}A$	A	A becomes 1 or remains at 1
$\Delta A + \nabla A + \overline{\Delta}A + \overline{\nabla}A$	1	All possible events for A
$\Delta A(\nabla A + \overline{\nabla}A + \overline{\Delta}A)$	0	'Anding' differing events for A

Using these bridging rules it is possible to describe the transition behaviour of conventional logic gates. As an example, let us consider a 2-input NOR gate (inputs A and B, output Y). Its EDL equations can be derived as:

$$\nabla Y = \Delta A + \Delta B \qquad (3a)$$

and

$$\Delta Y = \nabla A \nabla B + \nabla A \overline{\Delta}B + \overline{\Delta}A \nabla B \qquad (3b)$$

As can be seen, the common combinational logic gates may not generate simple EDL functions. In transition based logic a different set of gates is used for the system design. Their definitions and implementation is described in the following section.

3. GAAS IMPLEMENTATION OF EDL GATES

The four basic gates required for the design of EDL systems are: the Muller-C element, the EDLINCOR, the EDLXOR and the inverter. A standard inverter can be regarded as an EDL element, although its function is slightly different from its conventional logic function. The inverter converts one transition of its input variable (eg. 0 to 1) into the other transition (1 to 0 for the example) at the output. It converts an event (non-event) at the input into an event (non-event) at the output. The EDLXOR function is such that any transition at one or the other of the inputs (but not both at the same time) produces a transition at the output. Both gates are the same as their conventional counterparts and, therefore, will not be discussed further here. Their DCFL implementation is shown in Figure 1.

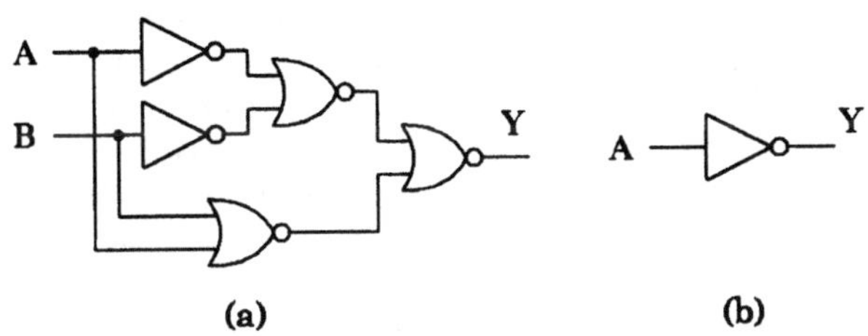

Figure 1. DCFL implementation of two EDL functions
(a) Exclusive-OR for events
(b) inverter as an EDL element

The Muller-C element and the EDLINCOR gate are sequential rather than combinational in their operation. The Muller-C element implements the AND function for events, such that if a specific transition takes place at one input and it is coincident with, or followed by, a similar transition of the other input(s), then that transition will be presented at the output [3]. In conventional logic terms its function can be described as:

$$Y(i+1) = Y(i)(A+B) + AB \tag{4}$$

A very efficient implementation of the Muller-C element can be derived using PDLL logic family primitives. Figure 2a shows a basic PDLL structure [5]. The gate consists of an input stage where the logic function is implemented, and a static latch stage which is constantly refreshing the internal node. The circuit is in a precharge phase when the clock signal is high and in an evaluation phase when the clock signal is low. Figure 2b shows the transition table for this circuit. It can be observed that if we use IN and $\overline{\Phi}$ as inputs, only the second entry in the table is different from the transition table of a Muller-C element. By controlling the clock input of the PDLL latch through a NOR gate and the pull-down transistor through an AND gate we obtain a circuit implementing the Muller-C gate, as shown in Figure 2c together with its transition table in Figure 2d.

The circuit can be further simplified by incorporating the AND function into the PDLL latch, since unlike DCFL, PDLL allows for the serial connection of transistors in the pull-down section. The final implementation of a static Muller-C element shown in Figure 3a. Associated HSPICE waveforms for the circuit operating at a signal frequency 1GHz using a 1V power supply are shown in Figure 3b. The layout of the Muller-C gate in 0.6μm HGaAs-III MESFET technology is shown in Figure 3d, where ring notation has been adopted in order to reduce the coupling between fast signal and power lines [6].

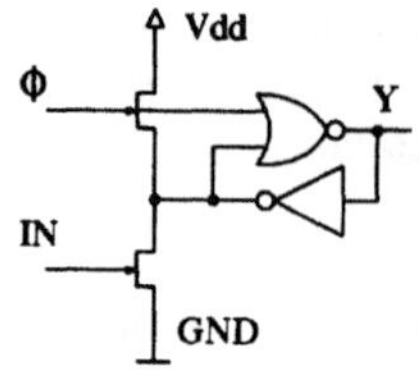

(a)

Φ	$\bar{\Phi}$	IN	Y(i+1)
1	0	0	0
1	0	1	0
0	1	0	Y(i)
0	1	1	1

(b)

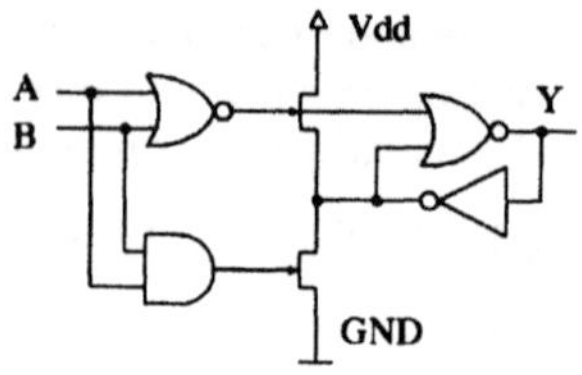

(c)

A	B	Y(i+1)
0	0	0
0	1	Y(i)
1	0	Y(i)
1	1	1

(d)

Figure2. Basic PDLL structure and its modification
(a) circuit diagram of a PDLL latch, (b) transition table of a PDLL latch,
(c) circuit diagram of a PDLL based Muller-C circuit,
(d) transition table of a Muller-C element

For comparison, the circuit schematic of the Muller-C gate implemented in standard DCFL together with its layout are shown in Figure 3c and Figure 3e, respectively. The PDLL-based circuit using MESFET technology consists of only 11 transistors. It dissipates 150μW of power which is less than half of the power required by a Muller-C using the standard DCFL approach. It also exhibits a smaller delay than DCFL. Two implementations have been compared. The first one is based upon 0.6μm MESFET HGaAs-III E/D process, and the second incorporates Quantum Well HEMT technology. The performance is compared using the figure of merit, η, representing the performance in terms of $(delay \cdot area \cdot power)^{-1}$. The performance comparison is shown in Table 2. As expected, the HEMT implementation is faster, but it dissipates more power and occupies more area than its MESFET counterpart.

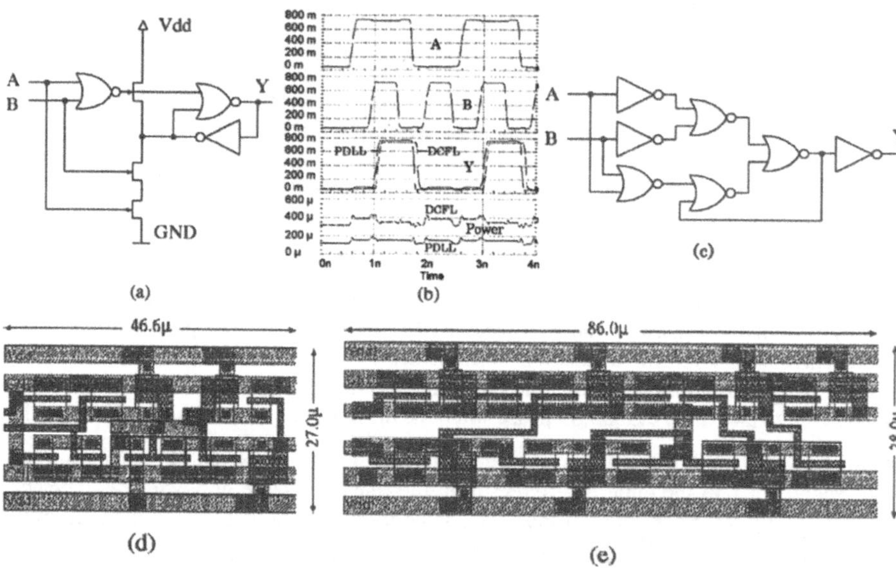

Figure 3. Muller-C element using PDLL primitives
(a) circuit diagram, (b) HSPICE simulation
(c) circuit diagram of the DCFL implementation
(d) layout of the circuit (a), (e) layout of the circuit (c)

Table 2. Performance comparison of the two input Muller-C element at $V_{DD} = 1V$

Technology/Family	No of transi-stors	Area (mm^2) $\cdot 10^{-3}$	Delay (ps)	Power dissipation (mW)	η $(ps \cdot mm^2 \cdot mW)^{-1}$
MESFET DCFL	18	2.408	120	0.35	9.888
MESFET PDLL	11	1.258	90	0.15	58.88
QW HEMT DCFL	18	9.100	85	5.2	0.2486
QW HEMT PDLL	11	4.900	65	2.1	1.495

The EDLINCOR gate implements the OR function for events, such that a transition of output Y is caused by a matching transition of either input A or input B or both simultaneously. For simultaneous transitions in opposite directions on A and B, Y retains its present value. Like Muller-C, this is a sequential circuit for which the input/output relation can be expressed in conventional logic terms as follows:

$$Y(i+1) = \overline{X}(i)(A+B) + AB \tag{5a}$$

where:

$$X(i+1) = X(i)(A+B) + AB \tag{5b}$$

The circuit diagram of the EDLINCOR using PDLL primitives is shown in Figure 4a together with associated HSPICE waveforms in Figure 4b, and the conventional DCFL implementation in Figure 4c. The PDLL implementation saves three transistors and as in the case of the Muller-C element, it is faster and exhibits lower power dissipation than the standard DCFL.

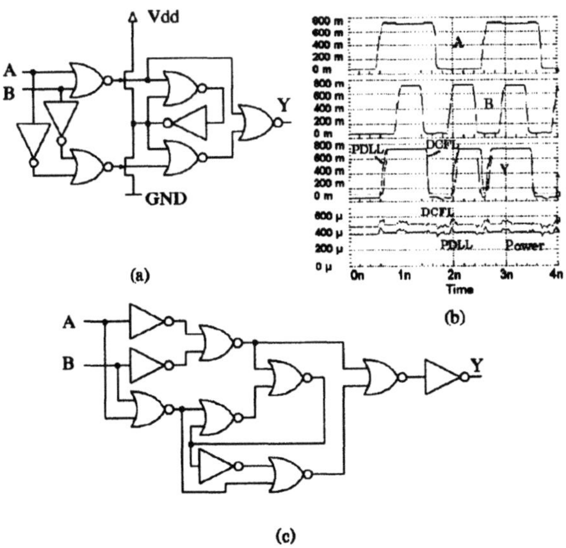

Figure 4. EDL INCLUSIVE-OR element using PDLL primitives
(a) circuit diagram
(b) HSPICE simulation
(c) circuit diagram of the DCFL implementation

4. CONCLUSIONS

A novel design of the two fundamental Event Driven Logic gates - the Muller-C element and the EDLINCOR - based on GaAs MESFET technology has been presented. The designs take the advantage of the excellent power-speed properties of the PDLL logic family. The circuits are compact, require smaller number of transistors and dissipate less power than a conventional DCFL implementation. They can be applied in the design of very high speed self-timed systems in Gallium Arsenide technology.

ACKNOWLEDGMENT

The support of the Australian Research Council and the Centre for Very High Speed Microelectronic Systems at Edith Cowan University is gratefully acknowledged.

REFERENCES

[1] W. Bowhill, *A 300 MHz Quad-Issue CMOS RISC Microprocessor*, Technical Digest of the 1995 ISSCC Conference, San Francisco, February 1995.

[2] N.R. Poole, *Self-timed logic circuits*, Electronics & Communication Engineering Journal, December 1994, pp. 261-270

[3] D.A. Pucknell, *Event-driven logic (EDL) approach to digital systems representation and related design processes*, IEE Proc. E, 1993, **140**, (2), pp. 119-126

[4] I.E. Sutherland, *Micropipelines*, Communications ACM, 1989, **38**, (6), pp. 720-738

[5] J.F. López, K. Eshraghian, R. Sarmiento and A. Núnez, *Gallium Arsenide pseudo-dynamic latched logic*, Electronics Letters, 1996, **32**, (15), pp. 1353-1355

[6] K. Eshraghian, R. Sarmiento, P.P. Carballo and A. Núnez, *Speed-area-power optimization for DCFL and SDCFL class of logic using ring notation*, Microprocess. Microprog., 1991, **32**, pp. 75-82

25

TRANSLATION OF C AND VHDL SPECIFICATIONS INTO INTERPRETED PETRI NETS FOR HARDWARE/SOFTWARE CODESIGN

Jaroslaw Mirkowski and Zbigniew Skowronski

Department of Computer Engineering and Electronics
Technical University of Zielona Góra
ul. Podgorna 50, 65-246 Zielona Gora
POLAND

ABSTRACT

Hardware/software codesign requires a formal model of designed system. Such a model should have several features, the most important of which are: to be well suited both for software and hardware representation, allow for different manipulations (including partitioning), and be able to cope explicitly with parallelism. Interpreted Petri nets can meet all three requirements. This paper outlines some practical aspects of transforming specifications given in C and VHDL into an interpreted Petri net.

1. INTRODUCTION

Managing heterogeneity in hardware/software codesign by the way of applying abstract models poses one of the greatest challenges in the whole codesign research [2]. While there exist many software- or hardware-oriented representations, it is hard to find satisfactory vehicle well suited for both. Although several approaches exist [7,2], each of them has some drawbacks.

The most popular intermediate format used is Control and Data Flow Graph, CDFG. CDFGs are main abstract models applied in high-level synthesis systems [2] and therefore were shifted to system-level synthesis. Apart from many advantages, CDFGs have one major drawback in case of codesign: events of the system can be expressed as linear order or lockstep of atomic operations. This limitation could be a severe restriction in case of heterogeneous systems, where partial order expressions are more natural or even necessary (due to unpredictable communication time, for example).

Other model widely used is communicating (co-operating) Finite State Machines (CFSMs) [2]. There exist different variations of this model (see [7] or [2] for a description). According to De Micheli [2] main drawback of applying CFSMs in codesign is a limitation of partitioning by input specification.

On the other hand Petri nets [6] appear to have all the necessary features for a good codesign model: strong theoretical background, explicit representation of parallelism, partial order of events, input specification independence and applications in both hardware and software domains. In order to apply them in practice it is necessary to translate typical specification languages (C and VHDL) into Petri Nets. Several approaches to this goal exist (compare [3]), but in case of VHDL the resulting net either does not implement in full the concept of simulation cycle or is not synthesizeable [5]. Here we are going to present an approach fully compliant with IEEE Standard VHDL Language Reference Manual (LRM) and suitable for further codesign tasks.

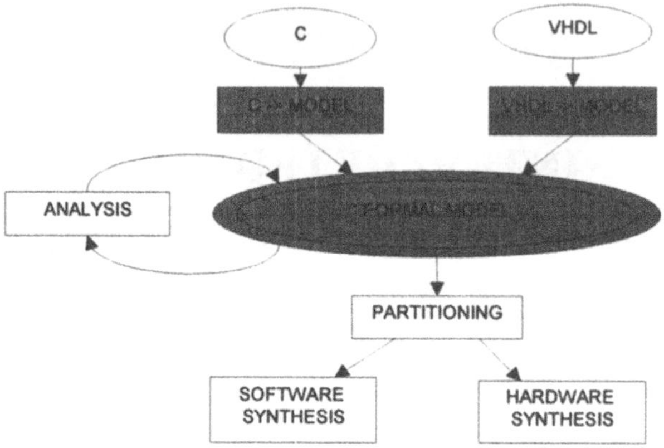

Figure 1. Simplified block diagram of a codesign operation flow.

2. HARDWARE/SOFTWARE CODESIGN

Hardware/Software Codesign [2,7] is an approach to the synthesis of fully integrated specification, analysis, and synthesis of systems based on microprocessors (realising the software part) and specialised hardware (Application Specific Integrated Circuits, ASIC) responsible for implementation of speed-critical parts of the system. The key point is the integrated approach to the whole system represented as one unit throughout most of the codesign process with the partitioning into software and hardware as the most important task. See the codesign tutorials [2] or [7] for details.

Although many different approaches to codesign exist, in general the design flow is as presented on Figure 1. The central point is the formal representation of the specified system. In our approach we have selected Petri nets for the reasons outlined in the Section 1. Shaded blocks on Figure 1 represent topics covered in this paper.

3. PETRI NETS

A Petri net is a bipartite, directed graph which has two types of nodes called *places*, represented by circles, and *transitions*, represented by bars or rectangles. Directed arcs connect the places and the transitions. A marking is an assignment of *tokens* (represented as black dots) to the places. The position and the number of tokens changes during the net execution according to *firing rules*. Formally, a Petri net *PN* is defined as a 4-tuple:

$$PN = (P,T,F,M_0)$$

where:

$P = \{p_1, p_2, \ldots, p_m)$ is a finite non-empty set of places

$T = \{t_1, t_2, \ldots, t_n\}$ is a finite non-empty set of transitions

$F \subseteq (P \times T) \cup (T \times P)$ is a finite non-empty set of arcs

$M_0: P \rightarrow \{0, 1, 2, \ldots\}$ is the initial marking

$P \cap T = \varnothing$ and $P \cup T \neq \varnothing$

A Petri Net for modelling hardware/software systems must fulfil following requirements:

1) It is *safe* (for every possible marking the number of tokens in every place is either 0 or 1);
2) It is *live* (during the net execution no transition may become unfireable on a permanent basis);
3) It is deterministic;
4) It is interpreted in the meaning as defined below.

An interpreted Petri Net exhibits the following three additional characteristics [1]: 1) It is synchronised (i.e. the firings of the transitions are synchronised on external events); 2) It is P-timed; in P-timed Petri nets a timing d_i, possibly of zero value, is associated with each place P_i. When a token is deposited in place P_i, this token must remain in this place at least for a time d_i. This token is said to be unavailable for this time. When the time d_i has elapsed, the token then becomes available. Only available tokens are considered for enabling conditions; 3) It comprises a data processing part whose state is defined by a set of variables $V=\{V_1, V_2, ...\}$. This state is modified by operations $O = \{O_1, O_2, ...\}$ which are associated with the places. An operation O_i assigned to a place P_i is carried out (executed) when P_i holds the token (is marked). The results of operations determine the value of the conditions (predicates) $C = \{C_1, C_2, ...\}$ which are associated with the transitions. A predicate may also describe some particular marking of the net. This is used especially in case of so called enabling and inhibitor arcs, which, unlike ordinary arcs, do not remove marking of a place while firing its output transition.

The net execution is performed according to the rules for transition enabling and firing [6]. A transition is enabled when all its input places handle a token. Enabled transition may fire when it is enabled and its predicate evaluates to TRUE. Firing a transition removes tokens from the transition's input places and places a token in every output place of this transition.

4. TRANSLATION FROM C TO IPN

Sequential operations comprising the sequential specification given in C are transformed into a Petri net together with parallelism extraction of the operations (preserving the semantics of the system).

Each operation given in the behavioural specification, notified as N_i, can be formally specified as a 6-tuple:

$$N_i = (In_i, Out_i, Op_i, Pr_i, Su_i, Con_i);$$

where:

In_i is a set of input variables of N_i;

Out_i is a set of output variables of N_i;

Op_i is a set of operands of N_i;

Pr_i is a set of predecessors of N_i;

Su_i is a set of successors of N_i;

Con_i is a predicate of N_i's execution.

Predecessors of each operation are determined through dependency analysis. Generally, an operation N_j is called dependent on N_k if N_k prepares data for N_j (data dependency) or its execution must be completed before N_j's execution can be started (control dependency). Typical situation in the latter case is when N_k is a conditional operation and the execution of N_j depends on the result of N_k, but does not use it as input data (like in the *if then else* construct).

Successors of an operation N_i are these operations, for which N_i is a predecessor.

Predicates, are logical conditions of executing operations. They are used to determine control dependencies and are reflected in the resulting Petri net. A predicate of an operation is a logical product of conditions that must be fulfilled to execute an operation. If an operation is not conditional, it is considered to have a predicate TRUE.

5. TRANSLATION FROM VHDL TO IPN

The compilation process of a VHDL behavioural specification into a Petri net consists of three main steps:

1. Transforming the operations of all processes into their respective IPN representation. This is performed in exactly the same way as in case of C specification described earlier.
2. Synchronisation of wait statements of all the processes.
3. Correct postponing of signal assignment (signal update).

The last two steps are related to simulation cycle of the VHDL language. Both the rules of simulation cycle and their influence for the sequence of operations are discussed in details elsewhere [5] and here only the consequences are presented.

Synchronisation of Processes. Any processes which exist in the VHDL specification have to be synchronised in such a way that no process can resume its execution until all other active processes suspend. Such a synchronisation can be introduced as a transition connecting all the wait statements. The method of the synchronisation fully follows the idea of Eles et al. [3] and is presented here for the completeness of the description.

Figure 2a presents a synchronisation of a wait statement (represented by the transition TW) in one process. The synchronisation transition connecting all the processes is denoted as TS. C is a predicate which is true when any of the signals from the wait sensitivity list changed during recent simulation cycle. The marking of the place NC denotes that next simulation cycle is started and this particular process is resumed.

Signal Update. A signal can be updated when two conditions are met: 1) Respective operation has been selected during the execution of the process (i.e. the process passed through the signal assignment operation); 2) The process resumed for the next simulation cycle (place NC from Figure 2 is marked).

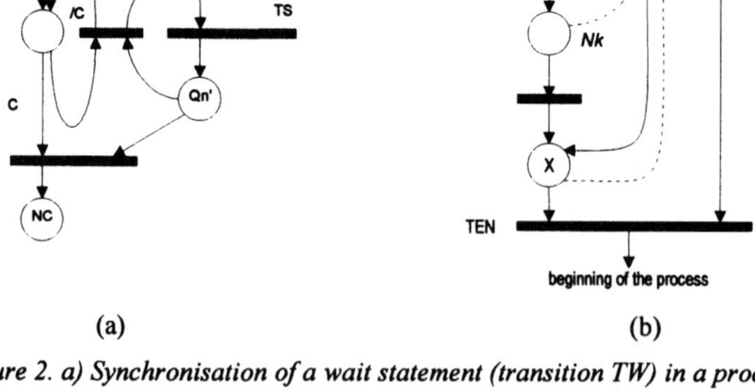

Figure 2. a) Synchronisation of a wait statement (transition TW) in a process; b) Postponement of signal update (operation N_k) and resuming the process

The correct location of a place representing signal assignment operation is determined by a parallelism extraction step described in section 3. Physical execution has to be postponed until the next simulation cycle begins. This is done in such a way that a place representing a signal assignment operation is substituted by a subnet (presented on Figure 2b) allowing for postponement of the physical assignment until resuming the process.

6. EXAMPLE

As an illustration of the concepts presented above a RS232 receiver specification taken from [8] will be used. It contains of two processes: *SyncRxIn* and *RS232.* The first of them is presented on Figure 3 and the IPN representing it is shown on Figure 4.

```
SyncRxIn: process (Clk, Reset_N)
  begin
    if Reset_N = '0' then          -- 1: C1:= (Reset_N='0')
      RxInSync <= '1';             -- 2
    elsif Clk'Event and Clk='1'    -- 3: C3:= (Clk'Event and Clk='1')
      then RxInSync <= RxIn;       -- 4
    end if;                        -- 5
  end process SyncRxIn;
```

Figure 3. An example of a process specification.

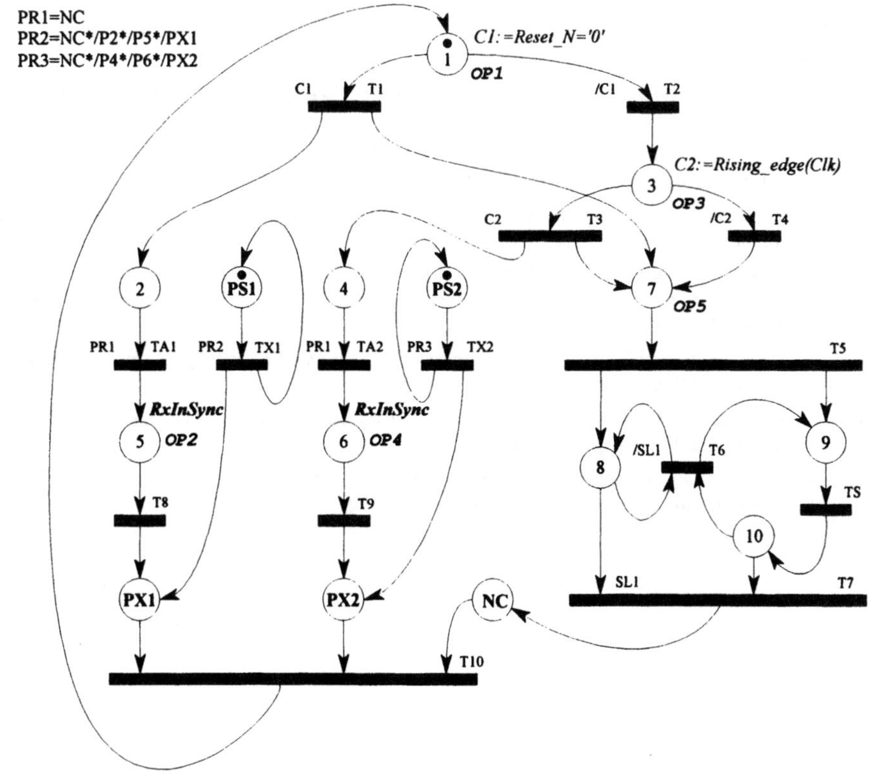

Figure 4. An interpreted Petri net for the VHDL process from Figure 3.

7. CONCLUSIONS AND FUTURE WORK

An idea of applying Synchronous Interpreted Petri Nets (SIPN) as a model for Hardware/Software Co-Design has been presented. As such this requires effective and correct-by-construction methods of translating input specifications given in widely used languages - C and/or VHDL. In the latter case, also LRM compliance is also very important in order to obtain the same simulation results from VDHL- and SIPN simulators. Such methods have been presented. The advantage of the VHDL translation method presented here over existing approach [3] is twofold:

- the consequences of simulation cycle (postponement of signal assignment) are introduced explicitly in the control structure of the net instead of implicit implementation inside the dataflow,
- possible parallelism of the operations is extracted and introduced during the translation, while in [3] processes are represented by sequential Petri subnets, which have to be parallelised in separate step.

The correctness of the VHDL-based approach has been verified manually. Main workflow now concentrates on implementation of both algorithms in order to automate the translation. Intensive research is also being done on the partitioning problem based on modelling results.

REFERENCES

[1] R.David and H. Alla, *Petri Nets for Modeling of Dynamic Systems - A Survey,* Automatica, vol. 30, No. 2, 1994, pp. 175-202.

[2] G.De Micheli, *Computer-Aided Hardware-Software Codesign,* IEEE Micro, vol. 14, No. 4, Aug. 1994, pp. 10-16.

[3] P.Eles, K. Kuchcinski, Z. Peng and M. Minea, *Synthesis of VHDL Concurrent Processes,* Proc. of the Euro-DAC, 1994, Grenoble, France, Spet. 19-23, 1994.

[4] D.D.Gajski and L. Ramachandran, *Introduction to High-Level Synthesis,* IEEE Design & Test of Computers, vol. 11, No. 4, Winter 1994, pp. 44-54.

[5] J.Mirkowski, K. Bilinski and E. Dagless, *Petri Net Modelling of VHDL Simulation Cycle for High Level Synthesis Purposes,* Proc. of the SIG-VHDL Spring'96 Working Conf. "VHDL-Forum for CAD in Europe", Dresden, Germany, May 5-8, 1996.

[6] T.Murata, *Petri Nets: Properties, Analysis and Applications,* Proceedings of the IEEE, vol. 77, No. 4, April 1989, pp. 548-580.

[7] W.Wolf, *Hardware/Software Co-Design of Embedded Systems,* Proceedings of the IEEE, vol. 82, No. 7, July 1994, pp. 967-989

[8] *VHDL Modelling Guidelines,* European Space Agency, European Space Research and Technology Centre; Report ASIC/001, September 1994.

26

FIPSOC. A NOVEL MIXED FPGA FOR SYSTEM PROTOTYPING

[1]J.M. Moreno, [1]J. Cabestany, [1]E. Cantó, [2]J. Faura, [3]P. van Duong, [4]M.A. Aguirre and [2]J.M. Insenser

[1]Universitat Politècnica de Catalunya, Gran Capitá s/n, Building C4
08034 Barcelona
SPAIN
(moreno@eel.upc.es)

[2]SIDSA, Parque Tecnológico de Madrid
28760 Tres Cantos (Madrid)
SPAIN
(faura@sidsa.es)

[3]MIKRON GmbH, Breslauer Strasse 1-3
85386 Eching
GERMANY
(pvduong@mikron.de)

[4]Universidad de Sevilla, Avda. Reina Mercedes s/n
41012 Sevilla
SPAIN
(aguirre@gte.esi.us.es)

ABSTRACT

In this paper we present a novel RAM-based field programmable mixed-signal integrated device consisting of a Field Programmable Gate Array (FPGA), a set of programmable and interconnectable analog cells, and a microprocessor core. This processor can run general purpose user programs, handle the dynamic reconfiguration of the programmable blocks and probe in real time internal digital and analog signals. The device is especially suitable for development and fast prototyping of mixed signal integrated applications.

1. INTRODUCTION

System designers have been craving for flexible prototyping systems onto which they could map large designs to validate them before fabrication. Typically, these designs may include a digital part, an analog part and a software program running on a microprocessor or microcontroller. However, these three domains (digital, analog and software) have to be designed and prototyped separately, using different CAD tools and hardware parts for each one.

Within this framework we introduce the FIPSOC (FIeld Programmable System On Chip) prototyping and integration system, consisting of a mixed-signal Field Programmable Device (FPD) with a standard 8051-microprocessor core, a suitable set of CAD tools to easily program it, and a set of library macros and cells which support a number of typical applications to be easily mapped onto the FPD and migrated to an ASIC afterwards, if required.

The advantage of this approach relies upon the fully integrated design and prototyping methodology that the user can follow with such a system, because he can download his application onto the programmable hardware and then use the internal microcontroller to probe it in real time (both digital and analog). A powerful integrated set of user-friendly CAD tools is provided. Also, a suitable library has been developed providing a very easy path for migration to ASIC after the prototyping phase.

We will focus on the FIPSOC chip architecture, and on the enhancements that an on-chip microprocessor can bring when included in a field programmable device. Following sections describe the circuit in its main parts, the multicontext dynamic reconfiguration possibility of this device and how it can be applied to hardware-software interaction. Then, we describe how this system can be used as a prototyping workbench with real time probing.

2. SYSTEM DESCRIPTION

The chip includes a Field Programmable Gate Array (FPGA), a set of fixed-functionality yet configurable analog cells, and a microprocessor core with RAM memory and some peripherals. The different interfaces between these blocks themselves and to the microprocessor provide a very powerful interaction between software, digital hardware and analog hardware. Figure 1 shows a block diagram of the FIPSOC device.

The chip has been designed using a full custom methodology for the FPGA and the analog area, and a synthesized *soft core* for the microcontroller. A 0.5μm triple metal layer CMOS 3V process provided by ATMEL ES2 was chosen to implement the first generation of this device.

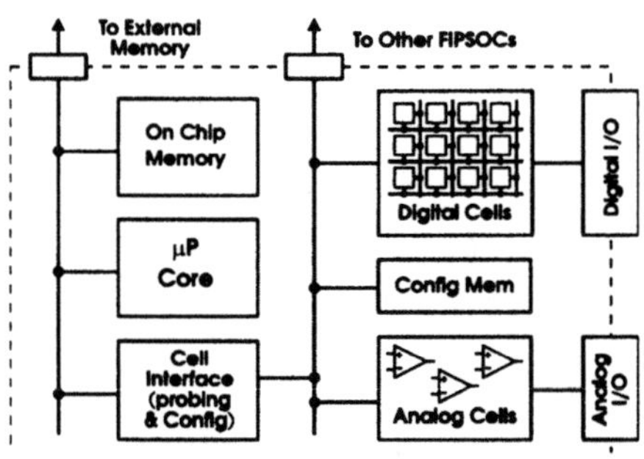

Figure 1. Block diagram of the FIPSOC chip

The FIPSOC device includes an array of programmable DMCs (Digital Macro Cell). The DMC is a large granularity, Look Up Table (LUT) based, synthesis targeted 4-bit wide programmable cell. Figure 2 shows a simplified block diagram of the DMC.

The DMC has two main blocks: a combinational part, composed of four 4-input LUTs, and a sequential block including four FFs. Between them there is an internal router which provides the necessary connectivity, and makes it possible to feed direct inputs into the FFs rather than using the combinational outputs. This makes it possible to use the combinational and the sequential blocks more or less independently. Each Look Up Table

(LUT) can implement any Boolean function of 4 inputs. Every two 4-input LUTs share two inputs, and two LUTs can be combined to form a 5 input function or a 4 to 1 multiplexer (four inputs and two control bits). The four LUTs of a DMC can be combined to perform any 6 input Boolean function. The whole combinational part of the DMC can be configured as a 16x4 RAM memory (in fact, two independent 16x2 memories) or as a cascadable 4-bit adder or subtractor with carry-in and carry-out (also some other arithmetic functions are possible).

The sequential part of the DMC includes four two-input flip-flops (FF), each of which can be independently configured as mux-type or enable-type, as latch or FF, and with synchronous and asynchronous set or reset. Again, the whole sequential part of the DMC can be configured as a cascadable shift register with load and enable or as a cascadable 4-bit up/down counter with load and enable.

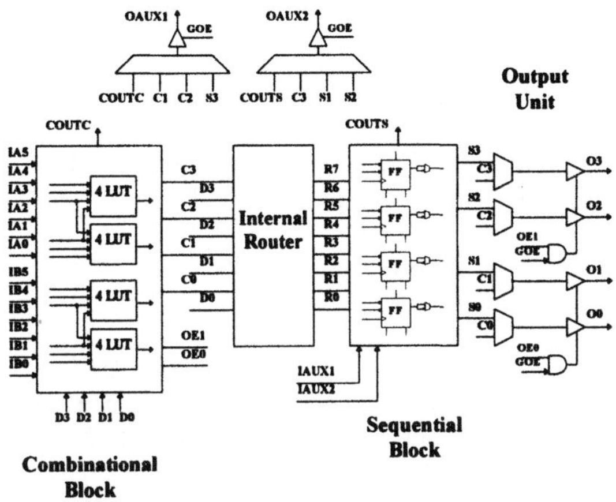

Figure 2. Simplified DMC block diagram

These combinational and sequential macro functions are especially suitable to be used by synthesis programs [1]. The routing architecture of this FPGA core has been designed according to this large granularity philosophy. Tracks spanning one, two and four DMCs (horizontally and vertically) are provided for general purpose interconnect. Long lines spanning the whole height or width of a column or a row and dedicated tracks for global reset and clock spine distribution are provided. Interconnection switches composed of MOS transistors are controlled with RAM memory cells writeable by the microprocessor.

The analog subsystem is composed of fixed functionality (yet programmable) blocks of coarse granularity. The basic building block is depicted in Figure 3. Each FIPSOC family members will have a different number of these blocks.

The analog block is intended to support four input/output analog channels with amplification, filtering, comparison and digital conversion. The functionality of the analog cells is fixed, although the cells themselves are programmable (i.e. the gain of the amplifiers or the accuracy of the DAC/ADC block can be selected).

A flexible interconnection architecture is provided to let the user build a custom application out of these blocks. In particular, nearly any internal point of the analog block can be routed to the ADC. Then, the microprocessor can use the ADC to probe in real time nearly any internal signal of the analog structure by dynamically reconfiguring these analog routing resources.

The ADC/DAC block is especially suitable for reconfigurable applications: it can be configured as one 10-bit DAC or ADC, two 9-bit DAC/ADCs, four 8-bit DAC/ADCs, or even one 9-bit DAC/ADC and two 8-bit DAC/ADCs at the same time. In the latter, two 8-bit DACs can be used to dynamically set the references for the 9-bit DAC/ADC, easily adjusting ranges and offsets on the fly.

As it has been indicated, the chip contains a standard 8051 microcontroller, which can be used either for general purpose user applications or for configuration tasks. This means that all commercial tools (assemblers, compilers, debuggers, etc.) available for 8051 can be used for the FIPSOC device.

Most of the power of this chip when used as a prototyping benchmark comes from the fact that the internal signals of the programmable hardware can be read and even sometimes written by the microprocessor as long as they are mapped as memory locations in the address map. For the digital part, the outputs of any DMC can be read as a memory location, and the FFs can also be written by the microprocessor in real time.

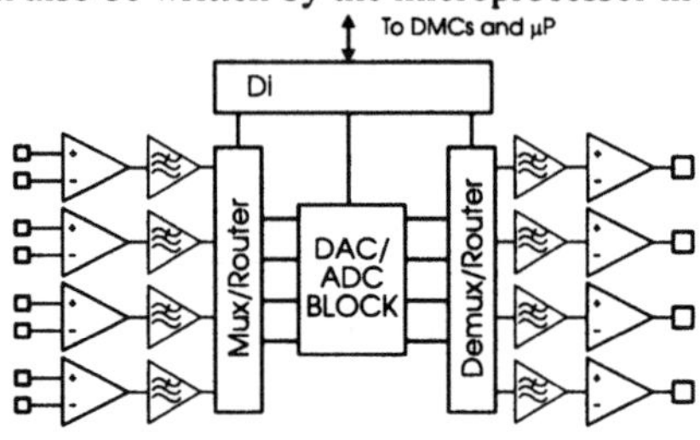

Figure 3. Analog block

For the analog blocks, the ADCs can be directly read and the DACs can be directly written by the microprocessor in real time, and the comparators can be read as memory locations as well. Finally, the microprocessor address bus can also be physically connected to the digital routing channels, which could be necessary for building microprocessor peripherals (for example communication ports, coprocessors, etc.). These communication points between the analog hardware and the digital world are also linked to the digital programmable hardware: The output of the comparators and the ADCs can be connected to the digital routing channels (and therefore to the DMCs inputs), and the inputs of the DACs can be driven by DMC outputs.

3. MULTICONTEXT DYNAMIC RECONFIGURATION

As it has been already mentioned, the chip configuration is managed by the internal microprocessor. To do so, the configuration memory is organized in words which are mapped onto the microprocessor address space. Furthermore, the configuration data are duplicated. It can be shown that such a duplication only needs a chip area overhead of below 12%. These two possible configurations are called *contexts*.

The microprocessor can then read and write these memory locations while in operation. This allows the user to reconfigure a context while the other one is still active, then change the active context to the new one. With this approach, the whole circuit can be reconfigured just by issuing a microprocessor command, and the reconfiguration time would be that of a microprocessor write cycle. In fact, a set of cells rather than the whole chip can be selected before applying the reconfiguration command. The main advantage of this technique is the possibility of loading the new context data *while* the active context is still in operation, thus not having to stop while the reconfiguration is taking place. Furthermore, the data inside the FFs are also duplicated, and can also be read and written by the microprocessor while the application is running.

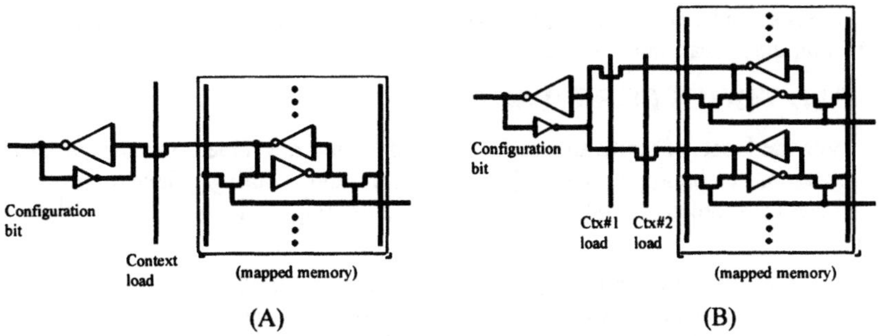

Figure 4. (A) One mapped context and one buffered one.
(B) Two mapped contexts and one buffered.

When the context is swapped, the status of the FFs can be maintained or stored with the rest of the context. This makes possible to initialize the FFs in the non-active context before setting it as active, and also to save the values of the circuit nodes when changing the context. Figure 4A shows this concept: The actual configuration bit is separated from the mapped memory through a NMOS switch. This switch can be used to load the information coming from the memory bus onto the configuration cell. This implementation is said to have one *mapped* context (mapped on the microprocessor memory space) and one *buffered* context (the *actual* configuration memory which directly drives the configuration signals). There exists also the possibility of having more than one mapped contexts to be transferred to the buffered context, like depicted in Figure 4B. However, as the number of mapped contexts increase, the efficiency of the proposed solution could decrease due to the bigger decoders needed to drive so much memory, and the size of the DMC would, of course, increase.

4. INTEGRATED PROTOTYPING WORKBENCH

Another interesting possibility comes from the optimized interface between the microprocessor and the programmable hardware itself, not its configuration memory as we have already studied. It is the possibility of probing in real time nearly any point of the analog or digital user application mapped onto the programmable hardware. In fact, it is possible to emulate a whole laboratory benchmark using just a FIPSOC Chip and a PC. The communication between them would normally be done through the on-chip RS232 serial port (an external driver is needed for RS232 voltage levels).

Logic data acquisition for this real time probing can be done in terms of memory read operations from the outputs of the DMCs, which are mapped as memory locations on the microprocessor memory space. LUTs configured as memories and dedicated DMCs configured as counters could also be used for fast logic data acquisition like in logic analizers. Note that the microprocessor can look up data from the LUTs while the LUT is in operation.

As it has been said, the internal ADC block can be dynamically rewired to probe nearly any internal point of the analog architecture. An analog data acquisition system, emulating a digital oscilloscope, could then be done using the microcontroller or some dedicated DMCs, as far as the digital side of the ADC can be connected to the digital routing channels and can be directly interfaced to the microprocessor. Even the internal DAC could also be used as a function generator, accepting data from the microprocessor core or from some DMCs configured as counters and memories.

The rest of the laboratory workbench is a matter of software: A digital oscilloscope could be emulated to present the acquired analog data (from the internal ADC) on the PC screen;

a logic state analyzer could be provided printing out the digital data obtained from the DMC outputs in real time.

5. CAD TOOLS FOR FIPSOC

An integrated set of software CAD tools is being developed for design entry and optimization, technology mapping, placement and routing, device programming, mixed-signal simulation and real-time system probing from a Windows™-based PC station.

A dynamic reconfiguration management tool, able to handle the multicontext operation of the chip, would constitute a very interesting research area here. Such a tool could analyse HDL code to check the coincidence in time of the processes, their criticallity and their system requirements. Up to now, some experiences on dynamic reconfiguration software have already been reported [2].

The FIPSOC chip is especially suitable for hardware-software co-design techniques due to the flexible interfaces between the programmable hardware areas and the microprocessor core, which results in a very powerful hardware-software interaction. This interaction is enhanced mainly due to: *a)* Internal signals from the programmable hardware can be probed and read as memory locations from the microprocessor core (analog signals have to be converted with the ADC). *b)* The microprocessor can dynamically reprogram a piece of hardware by overwritting the configuration memory. A co-design CAD tool could then be targeted to this device, putting in hardware those critical processes needing a high computational speed, and performing with software those tasks which would be prohibitively area-consuming.

6. CONCLUSIONS

A new concept to mixed-signal system design and prototyping for the FIPSOC device has been described and is currently at prototype level. The key point is the integrated methodology that can be carried out due to the flexibility of the configurable analog and digital hardware, and the simple interface between the digital resources, the analog subsystem and the microprocessor. It is estimated that the design cycle can be cut down by 30-40% by the use of the configurable hardware and the integrated emulation and verification design flows, compared with the use of separate analog and digital off-the-shelf FPGAs with their corresponding design tools. The use of the FIPSOC chip entails immediate reduction of PCB space, device reusability, dynamic reconfigurability and small time-to-market, which altogether makes the chip more than suitable for prototyping, pre-series fabrication and microelectronics research.

ACKNOWLEDGEMENTS

This work is being carried out under the ESPRIT project 21625. The authors would like to thank the European Commission for the financial support. The work has also the support of the Spanish CICYT under contract TIC96-2015-CE.

REFERENCES

[1] A.Stansfield and I.Page, *The design of a new FPGA architecture*, European FPL'95, Oxford (UK).

[2] P.Lysaght and J.Stockwood, *A Simulation Tool for Dynamically Reconfigurable Field Programmable Gate Arrays*, IEEE Trans. on VLSI Systems, Vol.4, n.3, Sep. 1996

27

A NORDIC PROJECT ON HIGH SPEED LOW POWER DESIGN IN SUB-MICRON CMOS TECHNOLOGY FOR MOBILE PHONES

Ole Olesen

Technical University of Denmark
Center for Integrated Electronics
Building 344, DK - 2800 Lyngby
DENMARK
Olesen@it.dtu.dk

ABSTRACT

This paper is a survey paper presenting the Nordic CONFRONT project and reporting some results from the group at CIE/DTU, Denmark. The objective of the project is to demonstrate the feasibility of sub-micron CMOS for the realisation of RF front-end circuits operating at frequencies in the 1.8-2.0GHz range.

The ultimate goal is a single-chip transceiver, requiring only an external band-pass filter between the chip and the antenna. DECT has been chosen as a comparative standard to compare the new approaches developed in the work as well as to facilitate good knowledge transfer to industry.

All circuit design is based on state-of-the-art CMOS technology (0.5μm and below) including circuits operating at 2GHz. CMOS technology is chosen, since a CMOS implementation is likely to be significantly cheaper than a bipolar or a BiCMOS solution, and it offers the possibility to integrate the predominantly digital base-band processing on the same chip.

Presently, only few examples of CMOS used for RF front-end circuits have been presented by academia, and so far no commercial products exist. The approach has been to do a CMOS block by block replacement of the blocks in traditional transceiver architectures. We feel that this approach is not ideal, since the excellent sampling properties of CMOS are not utilised. Therefore, the work focuses on developing transceiver architectures and circuits that truly exploit the desirable properties. At the same time, the work will investigate the possibility of including good off-chip components in the design by use of innovative, inexpensive package technology.

To achieve a higher level of integration, the project will use a novel codesign approach to the design strategy. Rather than making specifications based on a purely architectural approach, the work uses a concurrent approach, where circuit designers and architecture designers co-operate on the design and specifications. This allows more circuit issues to be included in the overall architecture, hopefully resulting in an architecture with circuit blocks suited for full integration.

1. INTRODUCTION TO CONFRONT

Concurrent Architecture and Circuit Development for a CMOS Dual-mode Wireless Front-end

The project is carried out by a consortium comprising the Technical University of Denmark (DTU, DK), Helsinki University of Technology (HUT, SF), and Royal Institute of Technology (KTH, S). These acronyms will be used in the following.

Each partner is in contact with well-known companies in each of the 3 countries.

Due to the complexity of the ambitious goals, around 12 full-time people are working on the project. Funding is national from each country.

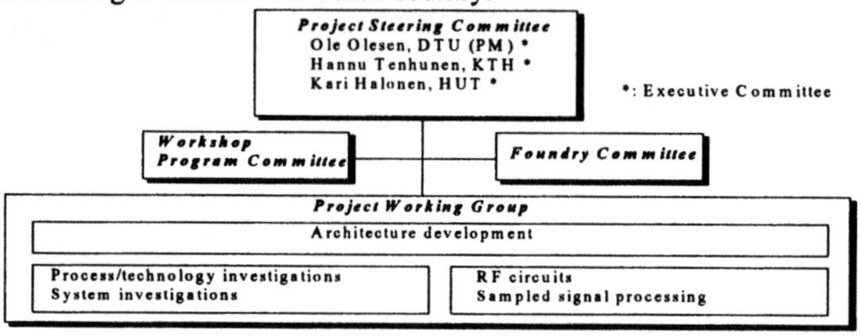

A part of the work is to be an investigation of the possibilities to implement a dual-mode system using the given circuits and architecture. DECT has been chosen as one of the standards, because the system is fully specified, and it is intended as a low cost system. Furthermore, DECT is capable of fairly high data rates (~1Mbit/s) required in future wireless systems (Multimedia ready systems). The specification is used at a comparative level to demonstrate the developed system's capabilities as a DECT system and as a dual-mode system. Dual-mode operation is of significant interest in the future where several wireless systems are likely to coexist.

Rather than developing a fully functional dual-mode DECT system, we propose to develop a test bench consisting of a combination of software and hardware solutions where the system's performance can be determined. The software part of the system is developed at an early stage in the work using Cadence Spectre's RF and HDL language. Part of this work is based on already existing knowledge [1] [2] [3] [4] [5] [6] [7]. The test bench allows not only an ongoing test of the system, but it also allows a mixed level test/simulation of the system where parts of the system can be described at a high level using the HDL, and other parts can be described by transistor net lists.

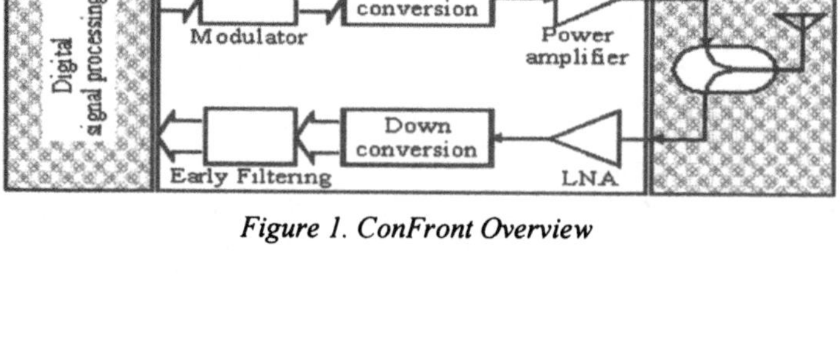

Figure 1. ConFront Overview

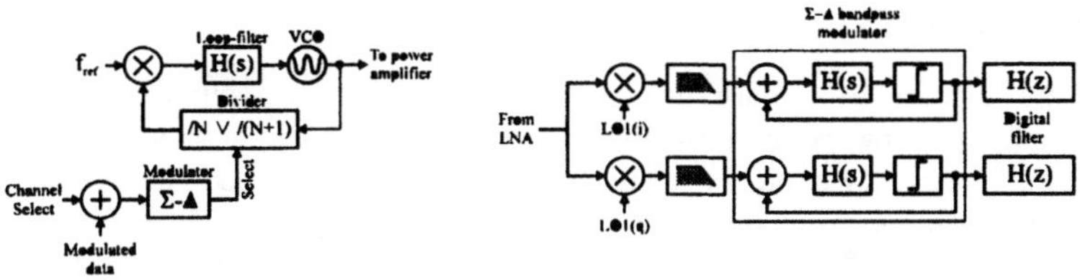

Figure 2. Alternative front-end architectures with emphasis on sampling properties

The nature of the work involves many critical issues. A general but very important obstacle is not to fall back to using a traditional architecture approach. Traditional architectures were developed to utilise properties of the traditional devices. Therefore, it is of vital importance that we look beyond traditional approaches to develop architectures better suited for CMOS.

However, at this stage several ideas for new approaches have already come up, and by pursuing these ideas, we expect to be able to fulfil the above conditions. CMOS has very desirable properties when designing sampled systems, so by moving the sampling process up in frequency, many traditionally analogue circuits can be replaced by digital signal processing.

2. RESULTS

In the following a few preliminary results from DTU are presented, but the project is in fast progress with more than 10 new publications already.

2.1. On-chip Inductors

Even though inductors are not widely considered as an option for CMOS designs, it is believed that on-chip inductors greatly improve the RF design options. Therefore, different approaches to implement inductors have been investigated [8]. These include the use of bonding wires as well as on-chip spiral inductors using several layers of metal to improve the inductor's quality factor. Focus is on inductors that can be designed with high reliability and yield in standard CMOS sub-micron processes. The modelling of on-chip components require access to an electromagnetic simulator like Sonet-Em or HP-Octave. Special focus will be given to modelling of the substrate, as significant improvements of the inductor's performance can be obtained here. The goal is to introduce a simple model to predict a given inductor layout with a reasonable precision.

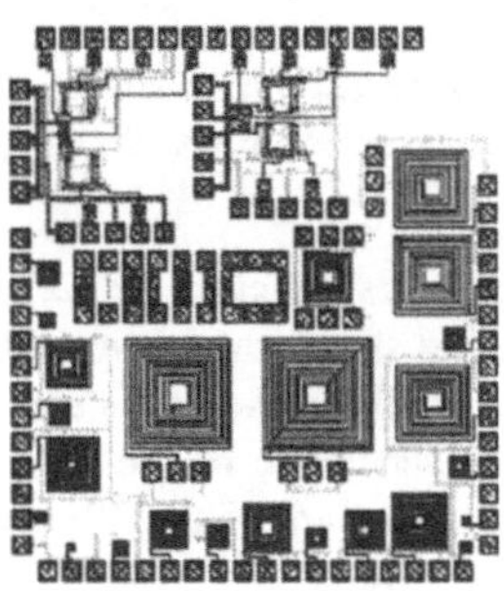

Figure 3. Spiral Inductor and VC Test Chip [8]

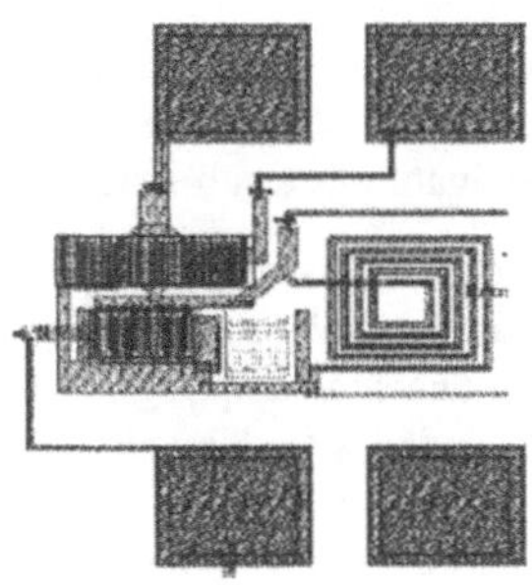

Figure 4. LC-Oscillator

Three test chips have been designed to investigate the feasibility of on-chip inductors for RF applications. Measurement show Q-values in the range 1-6.

2.2. VCO Design

It is likely that the VCO must include quadrature output, so different approaches to implementation of low phase noise VCOs are investigated. Furthermore, the VCO must have a broad frequency range due to the requirements for dual-mode operation (different carrier frequencies) and due to temperature and process variations. Most likely the VCO utilises high quality off-chip components to improve its phase noise performance if it is concluded that the use of off-chip components is a viable solution. However, fully integrated LC oscillators and other VCO structures are also being investigated [9,10,11,12, 13,14,15,16,17,18,19]. A fully integrated LC-oscillator is shown in Figure 4. Furthermore, high speed quadrature ring oscillators are also investigated.

Measurements from a test die show a free running frequency of 1.86 GHz and low phase noise. Measurements showed agreement with simulation.

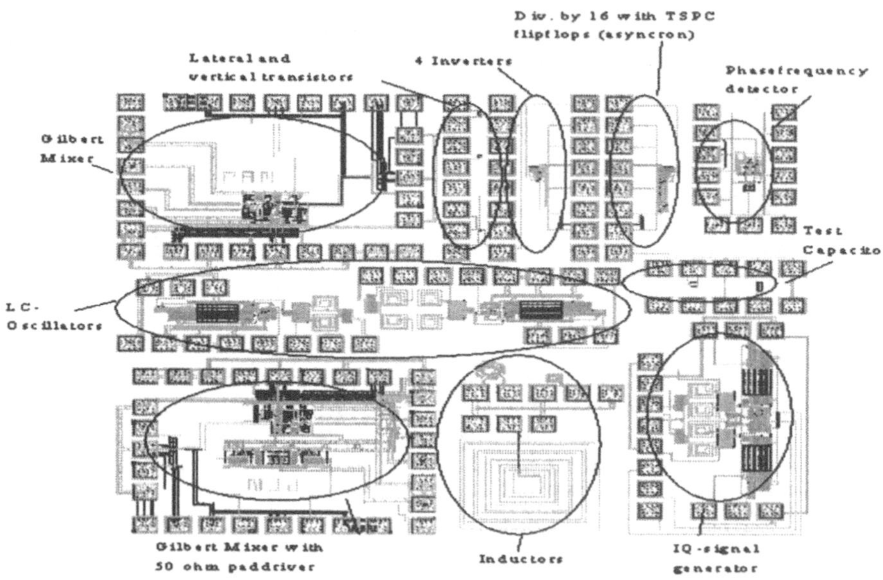

Figure 5. Test chips like this in different CMOS technologies

2.3. Mixer Designs

Design of mixers for up and down conversion. Different mixer architectures must be considered based on the selected architecture. A mixer topology for down conversion worth investigating is a sub-sampler.

Figure 5 shows a sub-sampling mixer with 250 transistors designed for a 2 GHz RF carrier frequency and sampling frequency of 10 Mhz using the Alcatel Mietec 0.5μm CMOS technology. The sub-sampling is carried out differentially in the first stage and converted to a single ended signal by the second stage. The schematic of the mixer is shown in Figure 6. Only two switches and two capacitors form part of the RF signal path and determine the cut off frequency and the sampling aperture. The required settling time of the OP-Amp depends only on the sampling frequency.

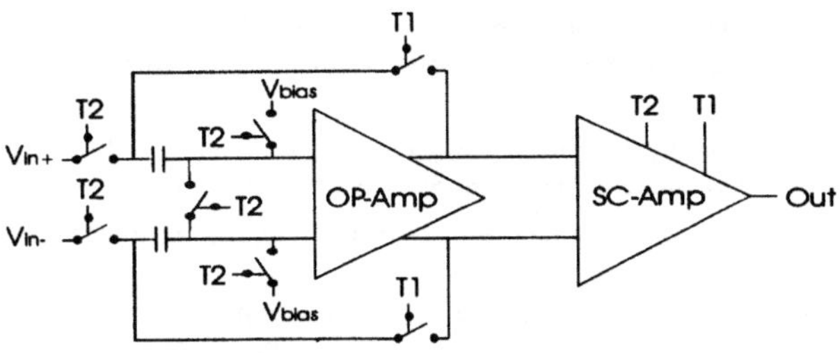

Figure 6. Schematic of Sub-sampling down conversion mixer

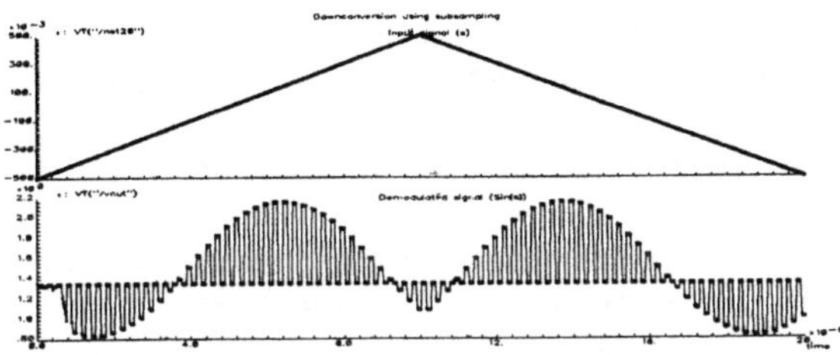

Figure 7. Simulation of Mixer performance

Figure 7 shows a simulation result of the mixer with a 2 GHz phase modulator carrier and 10MHz sampling frequency.

3. CONCLUSIONS

The paper has presented some of the first results of a large Nordic co-operation project started in 1996 but based on many years of research in the RF-field. The goal of the project is to develop new architectures in a codesign process using the advantages of deep sub-micron CMOS technology. One-chip-solution is the goal. The purpose of the paper is to give status and a survey of some preliminary results. The total project is ambitious and a few other research groups are working in the same field. Some detailed results have been presented already as mentioned in the references and 10 more detailed papers have been written.

The preliminary measurements on inductors and VCO on chip are promising. The hope is that this gives new ways in RF design for mobile phones, which is an area where Nordic industry has heavy positions, and a field with a glorious future ahead. Even applications in other fields are proposed.

ACKNOWLEDGEMENTS

The author would like to thank Professor Hannu Tenhunen, KTH, Stockholm, Professor Kari Halonen, HUT Helsinki, Director Ivan Ring Nielsen, Technoconsult, Dr Christian Řlgaard, National Semiconductors and staff members at DTU working on the project for excellent co-operation. The support from EU DGIII for foundry software and training has been invaluable for this project. The funding for the Danish part of the project has been granted by the National Technical Research Counsel.

REFERENCES

[1] Jacob Midtgaard and Christer Svensson, *5.8Gbit/s 16:1 Multiplexer and 1:16 Demultiplexer Using 1.2 μ BiCMOS*, Proceedings of ISCAS '94, London, 1994.

[2] Christian Řlgaard and A. Roforougan, *A Low Power 900 Mhz Tuned CMOS Amplifier with Large Output Swing Capability*, Proceedings of the 11th NORCHIP seminar, pp 162-169, Trondheim, 1993.

[3] Roforougan, A. Roforougan, Christian Řlgaard and A. Abidi, *A 900 Mhz CMOS RF Power Amplifier with Programable output*, VLSI Symposium, Hawaii, 1994.

[4] Christian Volf Řlgaard, *Analog Interface Circuits*, Ph.D. thesis, Department of Computer Science, Technical University of Denmark, 1995.

[5] Jacob Midtgaard, *Design Methods and Techniques for High-Speed VLSI-Circuits*,, Ph.D. thesis, Department of Computer Science, Technical University of Denmark, 1995.

[6] Christian Volf Řlgaard and Ivan Riis Nielsen, *Noise Improvement beyond the kT/C Limit in Low Frequency C-T Filters*, IEEE Transaction on Circuits and Systems II, Analog and Digital Processing, Vol. 43, no. 8, pp 560-569, August 1996.

[7] Christian Volf Řlgaard, D. H. Sassene, and Ivan Riis Nielsen, *An area efficient low 100 Hz low pass filter*, Proceedings International Symposium on Circuits and Systems, Vol. 1, pp 277-289, May 1996.

[8] Carsten Fallesen, *A 1.8 GHz Voltage Controlled Oscillator*, Proceedings NorChip'96, Helsinki, November 1996 .

[9] Nguyen and R. G. Meyer, *A 1.8-GHz monolithic LC voltage-controlled oscillator*, in IEEE Journal of Solid-State Circuits, vol. 27, pp. 444-450, March 1992.

[10] Basedau and Q. Huang, *A 1-GHz, 1.5-V monolithic LC oscillator in 1-μm CMOS*, in Proc. of the 1994 European Solid-State Circuits Conference, Ulm, pp. 172-175, Sept. 1994.

[11] Craninckx and M. Steyart, *A 1.8-GHz low-phase-noise voltage controlled oscillator with prescaler*, in IEEE Journal of Solid-State Circuits, vol. 30, pp. 1474-1482, Dec. 1995.

[12] Soyeur, K. A. Jenkins, J. N. Burghartz, H. A. Ainspan, F. J. Canora, S. Ponnapalli, J. F. Ewen, and W. E. Pence, *A 2.4-GHz silicon bipolar oscillator with integrated resonator*, in IEEE Journal of Solid-State Circuits, vol. 31, pp. 268-270, Feb. 1996.

[13] Ali and L. Tham, *A 900-MHz frequency synthesizer with integrated LC voltage-controlled oscillato,r* in ISSCC Dig. of Tech. Papers, San Francisco, pp. 390-391, Feb. 1996.

[14] Rofougaran, J. Rael, M. Rofourgan, and A. Abidi, *A 900-MHz CMOS LC-oscillator with quadrature outputs*, in ISSCC Dig. of Tech. Papers, San Francisco, pp. 392-393, Feb. 1996.

[15] Soyeur, K. A. Jenkins, J. N. Burghartz, and M. D. Hulvey, *A 3-V 4-GHz nMOS voltage-controlled oscillator with integrated resonator*, in IEEE Journal of Solid-State Circuits, vol. 31, pp. 2042-2045, Dec. 1996.

[16] Craninckx and M. Steyart, *A 1.8-GHz low-phase noise spiral-LC CMOS VCO*, in 1996 Symposium on VLSI Circuits, Honolulu, pp. 30-31, June 1996.

[17] Razavi, *A 1.8-ghz CMOS voltage-controlled oscillator*, in ISSCC Dig. of Tech. Papers, San Francisco, p. 23.7, Feb. 1997.

[18] Dauphinee, M. Copeland, and P. Schvan, *A balanced 1.5-ghz voltage-controlled oscillator with an integrated LC-resonator*, in ISSCC Dig. of Tech. Papers, San Francisco, p. 23.7, Feb. 1997.

[19] Jansen, K. Negus, and D. Lee, *Silicon bipolar VCO family for 1.1 to 2.2GHz with fully integrated tank and tuning circuits*, in ISSCC Dig. of Tech. Papers, San Francisco, p. 23.8, Feb. 1997.

28

A REUSE CONCEPT FOR AN I^2C-BUS INTERFACE

M. Padeffke and W. Glauert

Friedrich Alexander University of Erlangen-Nuremberg
Institute for Computer Aided Circuit Design
Cauerstrasse 6, 91058 Erlangen
GERMANY
Phone: (x49) 9131 8586-90, Fax: -99
E-Mail: padeffke@lrs.e-technik.uni-erlangen.de

ABSTRACT

In this paper a reuse concept for an I^2C-Bus interface is described. The goal was to make a verified model and a corresponding testbench available which can be reused in several designs. Depending on the requirements of the application the I^2C-Bus interface can have different functional features. This has been implemented using a parameterizable and configurable VHDL design. The efficient test of the multifunctional VHDL design requires special attention. This has been achieved using a parameterizable and configurable testbench which is configured by an ASCII file. Also the simulation itself can be controlled by text files.

1. INTRODUCTION

The reuse of an existing design is made complicated by several problems. Additional effort is required for the design of a reusable module. Neverless are "Design Reuse" and "Multiuse" regarded as a major contribution to the improvement of design productivity especially in connection with VHDL models [1,2,3]. But there are open questions concerning the additional overhead. The cost to acquire a "reuse" model and the time required to find a suitable model, the effort to capture the functionality of the model and to verify it, and to understand the correct use of it lead to acceptance problems for reusable components. Even inhouse development of reusable models is subject to some of these problems. But because of the better access to Know How and better support for the user an inhouse solution can be attractive, especially for components where frequent reuse is to be expected.

This paper describes the concept for a reusable VHDL design. We created a parameterizable and configurable VHDL model of an I^2C-Bus interface and a corresponding testbench. Depending on the requirements of the intended use the I^2C-Bus interface can be instantiated with a selection of functional features which will be defined using parameters respective VHDL-generics. A very important aspect is the verification of the parameterizable model. It is necessary not only to verify the complete functionality but also to make the testbench easy to reuse for other designers, to verify the parameterizable interface model and to integrate the testbench into their own design specific testbench. Therefore the testbench itself is parameterizable and configurable. For this purpose it contains several modules for an automatic check of the functionality. The simulation can be controlled by ASCII files. Chapter 2 describes the basic function of the I^2C-Bus, chapter 3 the parameterizable and configurable I^2C-Bus interface and chapter 4 the testbench.

2. THE I^2C-BUS

The I^2C-Bus is primary intended to transmit control information between several ICs of a system [4]. It consists of the two bidirectional buslines SDA (serial data) and SCL (serial clock) used in a wired-AND configuration. The number of ICs connected to the bus is only physically limited by the maximal wire capacity allowed, which is 400 pF. Each participant is recognized by a unique address and can be a transmitter (sending data, e.g. memory) and/or a receiver (e.g. LCD driver). Further a device can be a master initiating data transfers or a slave as a "passive" device. If two masters initiate a data transfer at the same time an arbitration takes place, where the master wins who addresses the slave with the lower address (wired-AND). An arbitration between several masters eventually leads to a delayed bus clock because all try to generate the SCL-clock signal. This leads to a slower transmission but has no other consequences. A slave has to synchronize to the bus clock but may slow down the transfer by holding the SCL down. The extension of the standard mode is the fast mode, which allows a bus clock frequency up to 400 kHz (standard mode: up to 100 kHz) and additional 10-bit addressing instead of pure 7-bit addressing.

3. THE PARAMETERIZABLE I^2C-BUS INTERFACE

The design goal for the parameterizable I^2C-Bus interface was to enable the user to utilize all the possibilities of the I^2C-Bus protocol and to make the instantiation control easy. It was achieved employing parameterizable and configurable VHDL code [5].

The overall model consists of three main parts: The BUS_LINE_INTERFACE, the SLAVE_APPLICATION_INTERFACE and the MASTER_APPLICATION_INTERFACE. Each block manages the data transfer to the corresponding extern modules. The latter two are instantiated depending on the value of the functional parameters.

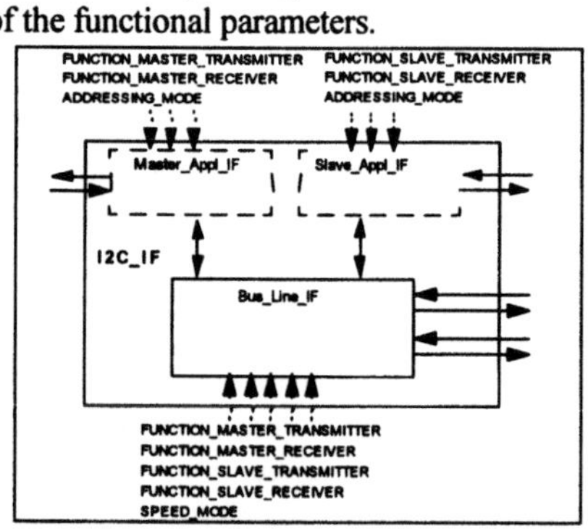

Figure 1. Parameterizable and configurable I^2C-Bus interface

Figure 1 shows the parameterization principle. The application interfaces are instantiated if at least one of the their function generics [7] (transmitter or receiver) is set to one, which has to be done before synthesis. While the function generics configure the application interfaces the address mode parameters determine the width of the address port. The third main module, the bus line interface, will always be instantiated. It evaluates all four functional parameters and also the speed parameter, which is responsible for the resulting SCL clock frequency, the resulting master interface will generate. With the described generics it is possible to implement different instantiations of the interface using this VHDL model. The following parameters are used [5]:

Functional parameters. Each function (Master, Slave, Transmitter, Receiver) can be implemented by setting the corresponding generic to 1. So in principle it is possible to implement a working interface which contains all four functions. The user has to take care to select a meaningful combination of these parameters. A master for example has to be always a transmitter, because he must be able to start the communication with a slave.

Addressing parameter. The addressing parameter determines wether a 10-bit address is allowed (1) or not (0). The parameter is valid for the master (target address) as well as for the slave (own address), if both functions (Master and Slave) are selected via the functional parameters.

Speed parameter. The I^{2}C-Bus spezification defines two different speed ranges:

- Standard mode: up to 100 kHz for the bus clock frequency
- Fast mode: up to 400 kHz for the bus clock frequency

For the reliable recognition of the transmitted data a digital filter is used and three samples must be taken within the minimal high time (600ns in fast mode). So the system clock for the interface has to have a minimal frequency of 5 MHz [(600ns/3)-1].

Table 1: Speed parameter

Mode	Code	(SCL) high:low	SDA-bit rate (kHz)	Clock frequency (Mhz)
Standard	0	25 : 25	100	5
	1	like mode 2		
Fast	2	7 : 12	263.2 ... 394.7	5 ... 7.5
	3	9 : 16	200 ... 400	5 ... 10

The above mentioned constraint leads to the speed parameters shown in Table 1. We planned to support four different modes but speed mode 1 was cancelled because of unsafe operations in some cases. The constraints result in a fixed system frequency for the standard mode. In the fast mode the system frequency may be between 5 and 10 MHz resulting in the different SDA-Bit rates, which can be achieved by the master interface.

With the parameter selection it is now possible to define the hardware which has to be implemented. The functional parameters control the instantiation of the corresponding modules with the help of generics and generate assignments [7]. All together there are 6 parameters which all can change the functionality of the system. The verification of the complete VHDL model is handled using a parameterizable testbench which is described in the next paragraph.

4. THE PARAMETERIZABLE TESTBENCH

The problem of having to test multiple scenarios for a parameterizable design requires an efficient test strategy. The following decisions were made:

Separation. The interface and the testbench were realized by two different designers. So a certain degree of independence was assured. The idea was to prevent making identical errors in reasoning. So a high rate of error detection is assured. Both designers revealed several errors in the designs.

Parameterization. A parameterizable design requires perforce a parameterizable testbench which is using the same parameter set as the design itself. Also the testbench becomes reusable and easy to adapt. All unneeded modules can be left out by setting the corresponding parameters. Figure 2 shows the modules used in the testbench ([5,6]). The three different application blocks each contain the I^2C-Bus interface and will only be instantiated if the corresponding functional parameter is set to one as for the interface itself. All unneeded ports can be ignored or driven with constant values. But the parameterization requires a consistency check, which is done in the testbench with assertion statements. They check not only the values of the original parameter but also specific consistency conditions, e.g. between the speed modes and clock periods.

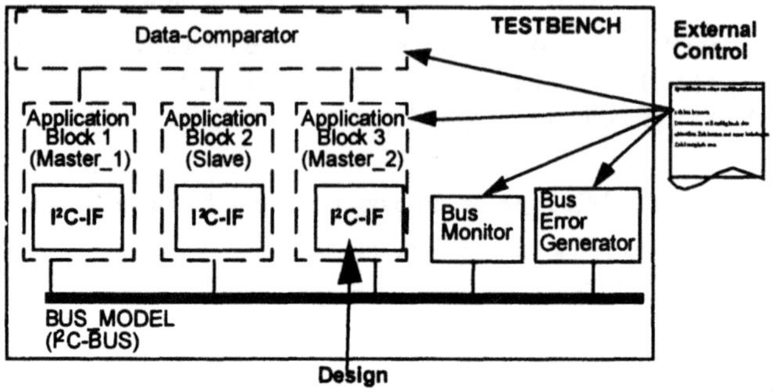

Figure 2. Block diagram of the testbench

Automation. For the effective verification of a parameterizable design a high degree of automatic verification is required. As there are no fully automatic verification tools we still have to simulate the VHDL design. But we can add modules to the testbench which take over some verification tasks like monitoring the bus or generating bus errors. The data_comparator for example compares always the received data with the ones sent from the transmitter. It reports every data mismatch. The comparator will only be instantiated if there are two or more application blocks instantiated. Because there is only one slave the data-comparator automatically multiplexes the correct parameter and signal values to the slave application block, so the winning master can communicate with the corresponding slave. The bus_model models the behaviour of the I^2C-Bus, but can be manipulated by the bus_error_generator. It can generate spikes, bitwise and bytewise errors or drive constant values on the bus wires. So the behaviour of the bus interfaces in case of an error can be tested. The bus_monitor checks whether the protocol is executed correctly. If not, an error message will be generated and eventually the simulation will be stopped. So the winning master can address the correct slave. The application blocks contain each an application model and the necessary bus interface.

Simulation. The modules of the testbench above proved to be sufficient to test all possible scenarios. The different modules help to verify the design by taking over several scenario specific tasks. The modularity of the testbench makes it easy to adapt to other design specific testbenches.

The parameter values of the I^2C-Bus interface are controlled by the entries in the parameter.dat file (see Figure 3). The parameter file is read at the beginning of the elaboration phase so only the modules are instantiated and the functionalities included which are needed for the current simulation run. Each application block has its own parameter row. In the first column a 1 indicates the instantiation of the module will be instantiated. The next four columns determine the values of the parameters also used in the design under test. With the last three columns several scenarios with completly different clock periods can be created. The last clock period for example determines the resolution of the BUS_MONITOR. Depending on this clock period the monitor is able to detect spikes of a minimal pulse width occuring on the two buslines.

1	1	1	1	2	200	51	50	MASTER_1
0	1	1	1	3	100	100	100	SLAVE
1	1	1	1	3	160	43	40	MASTER_2
Instantiation	T	R	A	SPEED	IF	APP	SIM	
0=No	R	E	D	_MODE	CLOCK	CLOCK	CLOCK	
1=Yes	A	C	R		PERIOD	PERIOD	PERIOD	

Figure 3. Example of a parameter file.

External Control. The use of files for parameterization and stimulation has several advantages.

- As a testbench has to be easy to use and to adapt it is helpful to have all parameters in one place (see Figure 3). So the parameters can be found and manipulated easily. A good commentation of the parameter file itself also helps the user.
- The stimuli files provide not only pure stimuli but together with the parameter file create scenarios needed to verify the complete protocol and the compatibility between different interfaces. So there may be for example two masters instantiated competing on the bus. But the user can inhibit one or both masters by letting them wait for a specific amount of time (Figure 4).
- While the stimuli file of a slave consists only of data to be transmitted on request the master stimuli file contains data either to be sent or to be received, the mode of the frame (see [4] and [6]), the slave address, the data to transmit and also the behaviour of the slave (that is how many bytes the slave will accept or send.). Together with the parameter file all possible scenarios may be built. The scenarios again can be changed easily by editing the stimuli file.
- The stimuli files of the slaves can be generated by former simulation runs or other tools, connected by text I/O.
- The response files can be analyzed after the simulation run.

Bytes	Dir	Mode	Addr	Stim	Wait	Slave	
4	0	01	104			1	First Frame
				1A			
				2A			
0	0				5000		Waittime in ns
0	1	01	108			5	Receive from 108 until X„FF„

Figure 4. Example of a master stimuli file.

In Figure 4 a section of a stimuli file for the master is shown. The entries in this file build the different scenarios of the simulation. The first column contains the number of bytes which will be sent in the current frame including the address bytes. The second column sets the MASTER_DATA_DIR signal and the third column the MASTER_MODE signal. The target address (hexadecimal) is written in the fourth column. If data have to be sent than in the following rows the fifth column contains the stimuli as hexadecimal numbers. An eventual wait time of the master is forced with an entry in the sixth column (see Figure 4). The last column contains the number of bytes which will be acknowledged and/or sent by the slave. This entry controls not only the byte numbers but also sets with the help of the data comparator the address of the slave interface to the target address in the stimuli file and the corresponding enable signals.

Modularity, good documentation and internal training in the Fraunhofer Gesellschaft (FhG-IIS) made the "reusable" I^2C_Bus interface accessible to designers. The interface has been integrated in two current designs already.

5. CONCLUSION

The paper shows an example of a parameterizable VHDL model including the corresponding testbench. The project demonstrates that there is not only some extra effort needed for the design of a parameterizable design but also for its verification. This leads to new test strategies, the main aspects of which are:

- Clear separation of the design under test and the testbench.
- Highest possible degree of automation for the verification. Therefore several modules have to be designed which realize this automation.
- Use of the same parameter set for the testbench and for the design under test
- Control of the simulation with files, including stimuli and response files.

Table 2. Design effort and synthesis results

	Interface	Testbench
No. of Files	28	15
No. of Lines	5180	4235
Design Effort	4 Months	3.5 Months

Optimization	Area	Time
Area	2492	45.83
Timing	3075	35.22
Compromise	2465	43.85

The question wether it is useful to design a parameterizable module instead of a rigid one can not be answered generally. The additional cost of Multi-Use and Reuse depends on several factors like design or purchase cost, ease of adaptation, verification of extern, may be licensed reuse components and so on. The example of the I^2C-Bus interface shows that a design of a parameterizable component is useful, but carries some overhead with it. Table 2 shows the extra effort of the design of a testbench for the I^2C-Bus interface. The hardware overhead compared to a standard I^2C-Bus interface could not be determined because there was no equivalent design for comparison.

Offering the model together with a testbench for reuse seems to be a good way to make the reuse acceptable for other designers. The model was successfully integrated in two VHDL designs.

REFERENCES

[1] N. Dutt, u.a., *Design-Reuse – Fact or Fiction.* 31st Design Automation Conference, San Diego/Californien, Juni 1994.

[2] V. Preis, R. Henftling, M. Schütz and S. März-Rössel, *A Reuse Scenario for the VHDL-Based Hardware Design Flow.* Proceedings of EURO-DAC with EURO-VHDL, Page 464 - 469, Brighton, September 1995.

[3] H.G. Büttner, *Setting up a Retrieval System for Design Reuse – Experiences and Acceptance,* Proceedings of EURO-DAC with EURO-VHDL, Page 575 - 578, Brighton, September 1995.

[4] *The I^2C-Bus and how to use it,* Paper from Philips, 1992.

[5] R. Kammerer and M. Padeffke, *Implementierung eines I^2C-Bus Interface Moduls,* internal Paper of FhG IIS-A, 1996.

[6] R. Kammerer and M. Padeffke, *Anwenderschnittselle des I^2C-Bus Interface Moduls,* internal Paper of FhG IIS-A, 1996.

[7] *IEEE Standard VHDL Language Reference Manual,* IEEE, 1988.

29

DYNAMIC ANALYSIS OF DIGITAL CIRCUITS WITH 5-VALUED SIMULATION

Raimund Ubar

Technical University of Tallinn
Ehitajate tee 5, EE0026, Tallinn
ESTONIA
Fax: (+372) 620 2253, E-mail: raiub@pld.ttu.ee

ABSTRACT

The paper presents a new method for 5-valued simulation of digital circuits based on calculation of Boolean derivatives on structural Binary Decision Diagrams - BDDs (or alternative graphs). The method is applicable for component level representations of digital circuits where as components arbitrary subcircuits (macros) instead of gates are considered. No dedicated model library of components for 5-valued simulation is needed. Instead of dedicated 5-valued models, generic ones in the form of structural BDDs are used. Advantages of the new approach compared to the traditional gate-level 5-valued simulation have been experimentally shown.

1. INTRODUCTION

Test pattern generation and fault simulation procedures for digital circuits, because of the need for high computational speed are usually based on simplified two-valued simulation. This results in the following:

- transitions between patterns in the test sequence can have transient pulses caused by hazards, which may result in the loss of credibility of test sequences in detecting faults of asynchronous circuits;
- it is also not possible to generate test patterns for faults in functionally redundant parts of the design and to analyse the quality of test sequences for delay faults.

The shortages of test design methods listed above can be overcome by using dynamic analysis methods based on multi-valued simulation. Multivalued simulation has been used for: detecting hazards in digital circuits [1], for delay fault analysis and test synthesis [2], for fault cover analysis and dynamic test generation [3] etc. In this approach, to each value from the given alphabet of signal values a special stylized waveform corresponds. The number of values (waveform types) can be different. Three-, five-, six-, eight-, nine-valued simulation alphabets are common. The drawbacks of traditional multi-valued simulation methods are the following:

- traditionally gate-level descriptions are used, which increases the complexity of the model and reduces the computational speed;
- circuits under test are represented usually by two-input gates;

- when, however, macroblocks are introduced, each of them should have its own computational model, which leads to the need of dedicated libraries.

In this paper for removing the listed drawbacks, a novel method for multivalued simulation is proposed, which lays on using Boolean differential calculus and structural BDDs (or alternative graphs) [4]. In the following, 5-valued simulation will be discussed.

2. MULTI-VALUED SIMULATION IN GATE-LEVEL CIRCUITS

For the purpose of dynamic analysis, the line waveforms are considered as members of the waveform-type set S = {0, 1, ε, h, x}. The members of S are described as follows: 0 (1) represents a type of waveform having a stable logic value 0 (1); ε (h) represents a waveform having a step-up transition from 0 to a final value of 1 (step-down transition from 1 to a final value of 0), and x represents unknown waveform. Let us call further 0,1 - static values, and ε, h, x - dynamic values.

Table 1.

OR	0 1 ε h x	AND	0 1 ε h x	NOT	
0	0 1 ε h x	0	0 0 0 0 0	0	1
1	1 1 1 1 1	1	0 1 ε h x	1	0
ε	ε 1 ε x x	ε	0 ε ε x x	ε	h
h	h 1 x h x	h	0 h x h x	h	ε
x	x 1 x x x	x	0 x x x x	x	x

Due to physical behaviors and the existence of delays in logic components, every line in the circuit can have one of the mentioned waveform-types of S. Correspondingly, the dynamic behavior of the circuit during one single transition period will be represented also by waveforms of S on the outputs of the circuit. Every gate in the circuit network can be regarded as an operator which computes the output value of the gate if the values on inputs are given. The operators for OR, AND and NOT gates in case of 5-valued simulation are depicted in Table 1.

From Table 1 and its transitivity we can compute the logic value of any line in the circuit which is represented as a network of two-input logic gates OR, AND or NOT.

3. FIVE-VALUED SIMULATION AND BOOLEAN DERIVATIVES

Let us represent a digital circuit by equivalent paranthesis form (EPF) synthesized by superposition procedure directly from the circuit. For synthesizing the EPF for a given circuit, numbers are first assigned to the gates and letters to the nets. Then, starting at an output and working back toward primary inputs, EPF replaces individual literals by products of literals or sums of literals. When an AND gate is encountered during backtracing, a product term is created in which the literals are the names of nets connected to the inputs of the AND gate. Encountering an OR gate causes a sum of literals to be formed, while encountering an inverter causes a literal to be complemented.

As an example the procedure is illustrated by transforming the circuit in Figure 1 to its equivalent paranthesis form:

$$\begin{aligned} Y &= (a_1\, b_1) = (c_{12} + d_{12})(m_{13} + e_{13}) = \\ &= (g_{124}\, h_{124} + f_{125}\, k_{125})(m_{13} + \neg k_{136}) = \\ &= (g_{124}\, h_{124} + \neg h_{1257}\, k_{125})(m_{13} + \neg k_{136}). \end{aligned}$$

When creating an equation by the superposition procedure described above, the identity of every signal path from the inputs to the outputs of the given circuit will be retained. Each literal in an EPF consists of a subscripted input variable or its complement which identifies a path from the variable to the output. From the manner in which the EPF is constructed, it can be seen that there will be at least one subscripted literal for every path from each input variable to the output. It is also easy to see that the complemented literals correspond to paths which contain an odd number of inversions.

Let us have an EPF $y = f(x_1, x_2, \ldots, x_i, \ldots, x_n)$ where x_i are literals, which describes the behaviour of a digital circuit. If a transition occurs on the input and affects the path denoted by x_i, then the transition will propagate up to the output y if $\partial y/\partial x_i = 1$, where $\partial y/\partial x^i$ is called partial Boolean derivative. In general case, if transitions occur on several inputs or a transition has a fan-out and propagates along several reconvergent paths, then the derivative $\partial y/\partial x_i$ may have a dynamic value $d \in VD$. As $\partial y/\partial x_i = d$ is the case, we are not allowed to exclude the possibility that during the transition a short period may exist where $d = 1$ happens. Hence, the statement made above can be generalized as follows: if a transition occurs on the input x_i, then the transition from x_i will propagate up to the output y if $\max\{\partial y/\partial x_i\} = 1$ is valid over the transition period. Now, the following two theorems can be easily proved.

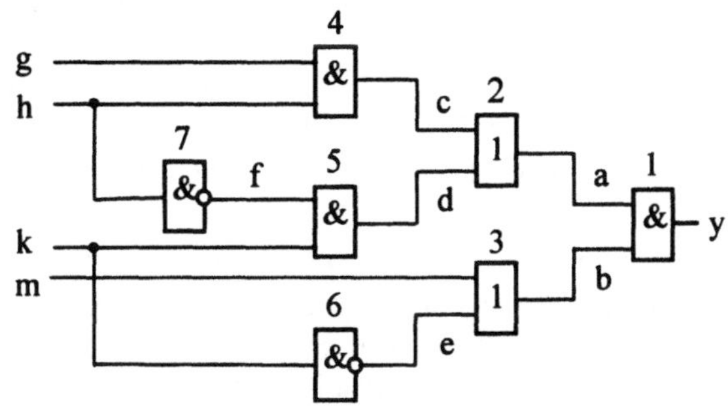

Figure 1. Digital circuit for creating the equivalent parenthesis form

Theorem 1

The value of the EPF $y = f(X) = f(x_1, x_2, \ldots, x_i, \ldots, x_n)$ for the given digital circuit in multivalued alphabet will be static if

$$\forall x_i \in X: \max_{x_j \in X_D}\{\partial y/\partial x_i\} = 0,$$

where $X_D \subseteq X$ is the set of literals whose values are dynamic.

Theorem 2

If $X_D \cap \{x_i \mid \max\{\partial y/\partial x_i\} = 1\} \neq \varnothing$ then the value of y can be calculated as the function of AND (or OR) of values of x_i for $x_i \in XD \cap \{x_i \mid \max\{\partial y/\partial x_i\} = 1\}$.

Proof

If the transition occurs on a single input x_i with $\max\{\partial y/\partial x_i\} = 1$ then $y = x_i$, i.e. the same transition (or its complement if x_i is inverted) occurs on the output. Otherwise, if several transitions on inputs x_i for $x_i \in X_D \cap \{x_i \mid \max\{\partial y/\partial x_i\} = 1\}$are propagating through the circuit, these paths can reconverge only on inputs of either AND or OR gates.

From Theorems 1 and 2 an algorithm can be developed for 5-valued simulation of digital circuits, which is based on calculating Boolean derivatives of EPFs. As the matter of fact, this algorithm obviously will be more complex compared to the traditional gate-level simulation based on multi-valued algebras. However, when using BDDs it is possible to create an efficient procedure for calculating derivatives.

4. BDDS AND EQUIVALENT BRACKET FORMS

As a general case of decision diagrams, alternative graphs (AG) [4] were proposed for representing digital systems. Unlike the traditional BDDs [5], structural BDDs (STBDD) or structural AGs (SAG) reported in [4] support structural representation of gate-level networks in terms of signal paths. By superposition procedure described in [4], we create STBDDs where one-to-one correspondence exists between graph nodes and signal paths in tree-like subciruits represented by STBDDs. We can consider a digital circuit as a network of tree-like subcircuits, each of them represented by an EPF. Consequently, a digital circuit can be represented by a system of STBDDs.Using STBDDs, it is possible to ascend from the gate-level descriptions of circuits to higher level descriptions without loosing accuracy of representing gate-level signal paths.

Denote the literal which labels a node m in a STBDD by x(m). We say that a value of the node variable activates the node output branch. According to the value of x(m), one of two output branches of m will be activated. A path in a BDD is called activated if all the branches that form this path are activated. The BDD is called activated to the value 0 (or 1) if there exists an activated path which includes both the root node and the terminal node labelled by the constant 0 (or 1). A STBDD G_y with nodes labelled by literals $x_1, x_2,..., x_n$, represents an EPF $y = f(X) = f(x_1, x_2,..., x_n)$, if for each pattern of X, the STBDD will be activated to the value which is equal to y.

As an example, Figure 2. shows a representation of a combinational circuit by a STBDD which corresponds to the EPF

$$y = (x_1\, x_{21} + \neg x_{22}\, x_{31})\,(x_{32}\, x_{51} + \neg x_4\, x_{61}) + x_{52} \neg x_{62}.$$

For simplicity, values of variables on branches are omitted (by convention, the right-hand branch corresponds to 1 and the lower-hand branch to 0). Also, terminal nodes with constants 0 and 1 are omitted (leaving the graph to the right corresponds to y = 1, and down - to y = 0).

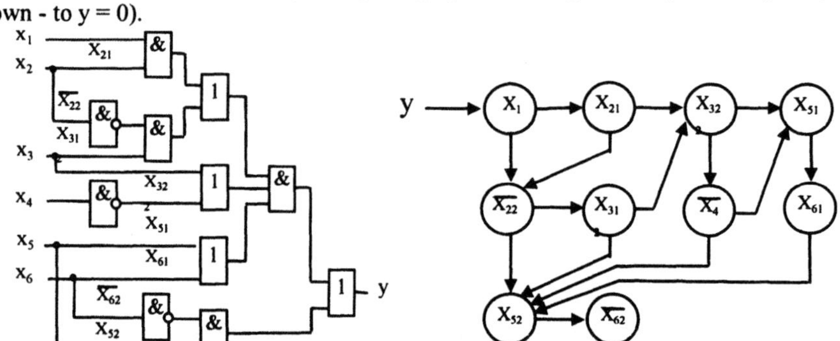

Figure 2. Combinational circuit and structural BDD

The graph contains 10 nodes, and each of them represents a signal path in a circuit (and a literal in the EPF). The literals in EPF and the related node variables in the graph correspond to input branches of the circuit in Figure 2.

5. FIVE-VALUED SIMULATION WITH STBDDS

Two-valued simulation on AGs is equivalent to path tracing on graphs according to the values of variables at a given test pattern. As a result of path tracing in G_y, the value of y will be calculated, which will be equal to the value of the label variable at the terminal node reached by path tracing. For explaining the calculation of Boolean derivatives, introduce the following notations: l(m) - activated path in the from the root node up to a node m; l(m,#1) (or l(m,#0)) - activated path from a node m up to the terminal node labelled by constant #1 (or #0); m^1 (or m^0) - successor of the node m for the value x(m)=1 (or x(m)=0). Denote l(m)=1 (or l(m,#e)=1) if there exists a path l(m) (or l(m,#e)), where $e \in \{0,1\}$; in other case, l(m)=0 (or l(m,#e)=0).

In the case of STBDDs, dy/dx(m)=1 is equivalent to one of the two conditions:

$$l(m) \wedge l(m^1,\#1) \wedge l(m^0,\#0) = 1 \quad (1)$$

$$l(m) \wedge l(m^1,\#0) \wedge l(m^0,\#1) = 1 \quad (2)$$

in other words, dy/dx(m)=1 is equivalent to the existence of simultaneously activated three paths: l(m), $l(m^1,\#1)$ (or $l(m^1,\#0)$) and $l(m^0,\#0)$ (or $l(m^0,\#1)$).

For example, for X=110011 in Figure 2, we have: $dy/\neg dx_4=1$, since the following paths are activated: $l(\neg x_4) = (x_1, x_{21}, x_{32}, \neg x_4)$, $l(x_{51},\#1) = (x_{51},\#1)$, $l(x_{52},\#0) = (x_{52},\neg x_{62}, \#0)$ and the condition (1) is fulfilled.

When calculating max{dy/dx(m)}, all dynamic values when tracing the path $l(m^1,\#1)$ should be taken as 1 and when tracing the path $l(m^1,\#1)$ as 0. When tracing l(m) all dynamic values should be taken either as 1 or as 0 properly so that the node m can be reached. In fact, instead of sequentially calculating the maximum derivatives separately step by step for all the nodes with dynamic values x(m), we can travers all the paths by a single procedure based on backtracking.

6. EXPERIMENTAL RESULTS

Dynamic test analysis has the goal to estimate the quality of test patterns by considering not only static levels of signals but also transitions between patterns in test sequences. The dynamic analysis package of the Turbo-Tester (TT) software [6] developed at the Tallinn Technical University is based on a 5-valued simulation on STBDDs. TT contains tools for multi-valued simulation, delay fault coverage analysis and dynamic fault analysis (especially oriented for detecting statically redundant faults). Experimental results on the multi-valued simulation of ISCAS'85 benchmark circuits are given in Table 2. Time ratio 1 shows the efficiency of macro-level simulation for the case when only a single signal transition at inputs is allowed, and Time ratio 2 corresponds to the case of multiple random transitions at inputs.

Table 2

Benchmark circuit	c432	c499	c880	c1355	c1908
Gate-level faults (G)	974	2194	1550	2194	2788
Macro faults (M)	616	1202	994	1618	1732
Fault ratio (M/G)	1,58	1,83	1,56	1,36	1,61
Time ratio 1 (M/G)	1,80	2,20	1,71	1,49	1,95
Time ratio 2 (M/G)	1,37	1,73	1,42	1,33	1,57

Benchmark circuit	c2670	c3540	c5315	c6288	c7552
Gate-level faults (G)	4150	5568	8638	9728	11590
Macro faults (M)	2626	3296	5424	7744	7104
Fault ratio (M/G)	1,58	1,69	1,59	1,26	1,63
Time ratio 1 (M/G)	1,91	2,36	2,05	1,43	2,07
Time ratio 2 (M/G)	1,44	1,87	1,64	1,15	1,97

Note that in Turbo-Tester, for the first time a macro-level multi-valued simulation method was implemented. As a result, the complexity of the model can be significally reduced compared with known methods which always need as a model a network of two-input logical gates. The fault ratio in Table 2 shows differences in model complexities. From Table 2 it follows that the speed of macro-level simulation based on the proposed method increases up to 2,36 times for given benchmark circuits compared to the gate-level simulation.

7. CONCLUSIONS

A new efficient 5-valued simulation approach for combinational or scan-based circuits for delay fault analysis, hazard detection or dynamic test analysis is presented. Its basic idea is substituting the traditional gate-level waveform calculation by nested Boolean differential calculus on structural BDDs. Introducing STBDDs allowed to reduce the complexity of the model by replacing low-level two-input-gate networks with higher macro-level representations. In fact, the multi-valued calculus on gate-level networks is transformed to a paths traversing procedure on BDDs. It is not needed to create for each new macro-block a separate dedicated multi-valued model. Instead, from the gate-level description automatically an STBDD-representation will be created, where a single general procedure for all types of macros will be used. Hence, no dedicated model library for multivalued simulation is needed. Experimental benchmark results have substantiated efficiency of the new approach compared to the traditional gate-level simulation approaches.

REFERENCES

[1] R. Andrew, *An algorithm for eight-valued simulation and hazard detection in gate networks*, Proc. of 16th Int. Symposium on Multiple Valued Logic, Blacksburg, 1986, pp. 273-280.

[2] W. Mao and M.D. Ciletti, *A variable observation method for testing delay faults*, Proc. Of 27th ACM/TEEE Design Automation Conference, 1990, pp. 728-731.

[3] S. Si, *Dynamic testing of redundant logic networks*, IEEE Trans. on Computers, 1978, Vol. C-27, No 9, pp. 828-832.

[4] R. Ubar, *Test Synthesis with Alternative Graphs*, IEEE Design & Test of Computers, Spring 1996, pp. 48-59.

[5] S. Akers, *Binary Decision Diagrams*, IEEE Trans. Computers, Vol. 27, No. 6, July 1978, pp. 509-516.

[6] R. Ubar, J. Raik, P. Paomets, E. Ivask, G. Jervan and A. Markus, *Low-Cost CAD System for Teaching Digital Test*, "Microelectronics Education", World Scientific Publishing, 1996, pp. 185-188.

PART V

Advanced Trends in Microelectronics Education

30

A FINITE STATE DESCRIPTION OF THE EARLIEST LOGICAL COMPUTER: THE JEVONS' MACHINE

Paul Amblard

Departement d'Informatique
Universite de GRENOBLE
LSR IMAG BP 53, F 38041 GRENOBLE Cedex 9
FRANCE
e.mail : Paul.Amblard@imag.fr

ABSTRACT

In 1870 a British scholar, W.S. Jevons, built the first computer designed to solve logical problems. Inspired by Boole and Babbage, Jevons built a mechanical device implementing automated deduction. In this paper, we present in contemporary terms the basic features of Jevons' machine. These features are related to finite-state machines and evaluation of words of a regular language. Two investigations about this machine are presented: the first one is evaluation of Jevons' machine complexity with a real-time environment. The second one is a VLSI implementation.

1. INTRODUCTION AND HISTORICAL ASPECTS

Homer, in the Iliad, used the word "automata" [3] to denote the gates of Heavens that opened without external action. The Greeks, Aristotle and others, established the rules of Logic. In the 18th century Leibniz tried to mechanize Logic but failed. In the 18h century Jacques Vaucanson, an engineer from Grenoble, built many highly complex mechanical automata. In the 19th century Boole established the formal rules of Logic. At the same time Babbage built the first mechanical calculating machine using stored "programs".

The last step towards mechanization of logic deduction remained to be done and it was accomplished by Jevons in 1870. [7] When he built his machine to solve syllogisms he had no idea that he was evaluating formulas of a regular language: of course, the theory of finite-state machines had not yet been established. One had to wait until the middle of the 20th century for its development.

In the first part of this paper Jevons' implementation is described using present day terminology: finite-state machine, regular languages, and so on.

The machine has the following interesting feature: the answer time is constant and instantaneous. Jevons' machine solves logical problems without delays. The price paid to attain this goal is a combinatorial explosion in space. If N is the number of variables, Jevons' machine has a number of states proportional to 2 at the power N.

In a second part of this paper we attempt to evaluate the exact number of states of Jevons' machine. It is a non-trivial exercise in minimization of automata. This is done using the Lustre language environment.

The third step sketches a VLSI implementation of Jevons' machine. This project has been implemented by students of our university.

2. THE MECHANICAL ASPECTS OF JEVONS' MACHINE

Jevons' machine deals with 4 logical variables. The hardware (more exactly the "woodware" because it was joinery) manages 16 wooden rods representing the 16 rows of the 4 variable truth table. The machine drives 16 evaluations in parallel. The machine looks like a small upright piano. A keyboard allows the user to introduce formula based on the classical syllogistic approach:

all A are B, all B and not C are D and not A, what could we say about the C ?

Cleverly designed levers and pins move the rods according to the keystrokes on the keyboard. The presentation of these operations is quite complex as we can see by this small paragraph quoted from Jevons' description.

> The full-stop key being now pressed has a double effect. [...] While a rod is in the *first position*, the lever passes between the pins and has no effect; but if the rod be lowered 1/2 inch into the *second position*, the lever will cause the rod to return to the *first position* by means of the alpha pin ; but if the rod be raised into the *third position*, the beta pin will come into gear, and the rod will be pushed 1/2 inch further into the *fourth position*.

The italic segments of the text were introduced by this author, they are not from Jevons himself. They make the description looking like that of an automaton.

3. JEVONS' MACHINE AS A FINITE-STATE MACHINE

We use present day language to describe what Jevons referred to as the "logical piano". The machine contains a keyboard like a piano, on which keys can be pressed to input a logical formula. One is allowed to hit one and only one key at a time.

So the set of keys constitute the terminal vocabulary and a sequence of keystrokes constitutes a word. The set of acceptable words is a language which we will see is regular. This means that a finite-state machine is sufficient to recognize if a given sequence of keystrokes is a member of the language or not. This syntactic verification was not done by Jevons. It should be obvious that the ideas about formal recognition of artificial languages date back from the second part of the 20th century.

We will first describe the language of Jevons formula, then the process of evaluation of a formula. A very interesting aspect is that syntactic analysis is driven by a finite automaton, and evaluation is also made by such automata.

3.1. The regular language of formula

We now describe a language of expressions close to those in Jevons' work by introducing some simplifications that do not change their expressive power. We present the syntax of the language and provide examples of the meaning of its "words". The language of formula is described by the following set of rules:

- a, b, c, d, a', b', c' and d' are the names of the variables where apostrophes denote the complementation of a variable. This set could be extended without altering the theory.
- Three other symbols are used: "." read as "dot", "+" read as "or", "=>" read as "implies". The symbols of operators have different priorities for evaluation.

The syntax is described by :

- boolean products using simple concatenation of variables

  examples: ab' bc' b'

  As usual, the complementation has priority over a product.

  Note that "aaa" is also a valid product (as in Jevons' machine, this is equivalent to "a").

- boolean sums of products using the "+" symbol

  examples: aa' + b + cd' ab d + a'

  As usual, the product has priority over a sum.

- implications using the "=>" symbol

  examples: a => b ab + c' => dc' + a ab => c+d

  (The implies sign corresponds in fact to a sum: a => b meaning not a or b)

  In an implication a => b, Jevons named subject the left part (in this case a) and predicate the right part (in this case b).

  a => b is interpreted as all A are B. We use this term subject throughout the paper.

- sequences of implications using the "." symbol.

  example: ab => c+d . d + a' => bc'

  A sequence of implications constitute a syllogism.

  All A are B and All B are C would appears as a => b . b => c .

  This composition by a dot has the meaning of an AND operation. Using a symbol for AND of variables (simple concatenation) and another one for AND of implications (dot) avoids the use of parentheses to denote priorities.

If we use a more standard boolean notation with the ' for NOT, concatenation for AND, + for OR, and parenthesis, the Jevons' expression

a b => c + d . d + a' => b c'

corresponds to

(a b => c + d) (d + a' => b c').

If we take into account that a => b is a' + b, we obtain the following formula:

((a b)' + (c + d)) ((d + a')' + (b c'))

The grammar of the above language can be expressed by the following rules where VT is the set of terminal symbols appearing in formula and VN is used by the metalanguage describing the language.

VT = { a, a', b, b', c, c', d, d', +, . , =>}

VN = {S, V, U}, where the axiom is S.

(S corresponds to a "Sentence", V to any Variable, U to a "sUbject")

S --> S V / S + V / U => V

U --> V / S . V / U V / U + V

V --> a / a' / b / b' / c / c' / d / d'

Note that we use --> for rewriting within grammar and => as a terminal symbol

The finite-state automaton accepting this language is given in Array 1. The symbol V denotes any Variable. State q0 is the initial state and state q3 is an accepting state. State q1 correspond to the recognition of an U. The "trap state" deals with all unaccepted words.

Array 1. Automaton recognizing Jevons' language.

```
next_state (q0,   V) = q1  ;next_state (q0,   +) = trap
                           ;next_state (q0,   .) = trap
                           ;next_state (q0, =>) = trap

next_state (q1,   V) = q1
next_state (q1,   +) = q0
next_state (q1, =>) = q2   ;next_state (q1,   .) = trap

next_state (q2,   V) = q3  ;next_state (q2,   +) = trap
                           ;next_state (q2,   .) = trap
                           ;next_state (q2, =>) = trap

next_state (q3,   V) = q3  ;next_state (q3, =>) = trap
next_state (q3,   +) = q2
next_state (q3,   .) = q0

next_state (trap, any input) = trap
```

3.2. The evaluation of a formula with boolean constants

We first examine the evaluation of a formula in the language. The first step is to consider the case of a formula with boolean constants True and False. The second step dealing with actual formula will be described in the next subsection.

Evaluation of

a b => c + d . d + a' => b c'

with constants a = False, b = True, c = False and d = True is evaluation of

0 1 => 0 + 1 . 1 + 1 => 1 1

(could be written False True => False + True . ...)

or (0 => 1) (1=>1) (This evaluation yields True)

The process of evaluation by an automaton is as follows:

> Consider a sub-language with only the constants True, False and a sign for AND. The syntax alternates boolean values (True or False) and an AND sign. Evaluation yields false as soon as one input constant is false. Evaluation is valid only at the end of a syntactically correct expression.

We can describe an evaluator with a Moore finite state machine with 4 states. It appears in Array 2.

> Initial state is q0. In q0 the expression is not yet correct. Evaluation gives nothing.
> In Trap state, the expression is definitively incorrect. Evaluation gives nothing.
> There is one state, named Good, where evaluation yields to True.
> There is one state, named Bad, where evaluation yields to False.

Array 2. Automaton evaluating product of constants.

```
next_state (q0,   True) = Good, next_state (q0, False) = Bad,
next_state (q0, AND) = Trap,
next_state (Good, AND) = q0, next_state (Good, True) = Trap
next_state (Good, False) = Trap
next_state (Bad, AND) = q0, next_state (Bad, True) = Trap
next_state (Bad, False) = Trap
next_state (Trap, any input) = Trap
output (Good) = True
output (Bad ) = False
output (q0 ) = undefined ; output (Trap) = undefined.
```

The evaluator of sums is similar and uses an OR. In this kind of boolean evaluators, the state actually represents the value of the result of evaluation.

The evaluator for implication is a bit more complex and has to deal with different values in subject and predicate part:

evaluation of an implication yields False if the subject is True and the predicate is False

Evaluation of a Jevons' formula combines

- one process for OR evaluation,
- one process for IMPLY evaluation and
- two processes for AND evaluation. (one for AND of variables and one for AND of sentences).

In Jevons' analysis, this combination of processes gives a 4 state automaton (Jevons did not deal with syntactical analysis, so equivalent states appeared).

The transitions are made upon recognition of the following terminals:

- The constants True or False, (the constant appears in a subject part or a predicate part of an implication). We use Subj and Subj' to denote the Subject part and the predicate part of an implication.
- The + sign appears in a subject part or a predicate part of an implication. We use +_subj and +_subj' to denote the + sign in the subject part or the predicate part.
- The => sign. It separates subject from predicate.
- The . (dot) sign. It separates implications.

The automaton of Array 3 is exactly the one described in textual form in Jevons' paper. Its initial state is labelled 1. The end of the evaluation process is marked by an input of a ".". The automaton then enters in either state 4 or state 1. In state 4 the value of the expression is considered to be False and in state 1 True. Once a False value is evaluated for an expression it remains False. This explains state 4 as a kind of trap state.

Array 3. Jevons' automaton to evaluate expressions with constants.

```
next_state (1, True ) = 1 ;next_state (1, => ) = 1 ;
next_state (1,    . ) = 1 ;
next_state (1,  False Subj') = 3 ;
next_state (1,       + Subj) = 3 ;
next_state (1,   False Subj) = 2 ;
next_state (1,      + Subj') = 2 ;

next_state (2,  + ) = 1; next_state (2,        . ) = 1;
next_state (2, =>) = 2; next_state (2, constant ) = 2;

next_state (3,         .) = 4 ;
next_state (3, => ) = 1; next_state (3, + subj' ) = 1;
next_state (3, constant) = 3;
next_state (3,+ subj) = 3;

next_state (4, any input ) = 4.

Output (1) = True;
Output (4) = False;
Output (2) = undefined ; Output (3) = undefined.
```

3.3. The general evaluation of Jevons' formulae

In this second step devoted to evaluation we deal with the complete evaluation, with variables. As an example let us consider the expression:

a b => c + d . d + a'=> bc'

Developing it as a product of canonical minterms gives:

a'bc'd'+ a'bc'd + ab'c'd' + abcd' + ab'cd' + abc'd

We could also minimize this expression, by Karnaugh maps for example, and obtain non canonical minterms. The expression is then a'bc' + ab'd' + acd' + bc'd. That expression, as a product of canonical minterms, can be considered as an "evaluation". It means that the expression is true if a is false and b is true and c is false and d is false or ...

This is the task of the Jevons' logical piano: giving the list of canonical minterms where the formula evaluates as "true". As in Jevons' machine, evaluation takes place in parallel: there are 16 evaluators, each corresponding to one row of the truth table where, obviously, the boolean variables are constants.

Let us describe the evaluation of the formula:

a b => c + d . d + a' => b c' .

on two different rows of the truth table, i.e. with different values of the variables a, b, c, d. For instance, for the combination a = False, b= False, c = False, d = True the expression is described by:

False_subj False_subj =>
False_subj' +_subj' True_subj' . True_subj +_subj True_subj =>
False_subj' True_subj' .

The sequence of states is: 2 2 2 2 2 2 1 1 3 3 1 3 3 4

The value is False.

With a = False, b = True, c = False and d = True, the expression

a b => c + d . d + a' => b c' .

is now

False_subj True_subj =>
False_subj' +_subj' True_subj' . True_subj +_subj True_subj =>
True_subj' True_subj' .

The sequence of states is 2 2 2 2 2 2 1 1 3 3 1 1 1 1

The result is True (as we have already seen in 3.2)

3.4. Evaluation of complexity

A first rough evaluation of complexity is as follows:

> for N variables, there are $L = 2^N$ rows in the truth table. There are L evaluators, and each one has 4 states.

So the global automaton for evaluation could have 4^L states. If we take syntactic analysis into account this number is multiplied by 4. Therefore in the case of 4 variables the global automaton including syntactic analysis and evaluation has $4 * 4^{16} = 2^{34}$ states.

As we shall see this is a really upper bound because the 16 evaluators do not work independently. The actual number of states is actually lower than 2^{34}. We have used Lustre environment to infer a lower bound for the complexity.

4. EVALUATING COMPLEXITY WITH THE LUSTRE ENVIRONMENT

LUSTRE is a language designed to describe real-time synchronous systems in a data-flow style [6]. We already used it to describe hardware, signal processing devices, etc. [1].

A program in Lustre may be seen as a set of constraints on the boolean outputs of a system in function of the boolean inputs. They are timed constraints and can make reference to internal boolean variables (or states). The compiler can establish the description of the automaton satisfying the specification given by the program (Lustre also deals with integers. We do not use this feature here. Of course integers do not give finite state machines).

This automaton is minimized by the software tools. The automaton can be translated to the C programming language. A current project translates it to VHDL for hardware synthesis.

We have done the experiment by coding the grammar of Jevons' language in Lustre equations. The Lustre compiler gave the 4 state automaton already seen about syntactical analysis. Then we directly transfered the four state automaton described by Jevons himself. We repeated it for each row of the truth table inserting the values of the corresponding variables.

4.1. Syntactic analysis

We took a small modification of the aforementioned grammar.

VT = { v, ., +, => }
VN = { S, U }
S --> U => v / S v / S + v
U --> v / S . v / U v / U + v

In the Lustre program

S . is noted Spoint, S + is noted Splus, U => is noted Uimpl, U + is noted Uplus.

At this level we do not distinguish the different variables : v is noted varia. The Lustre program of the syntactic recognizer is simply the following one:

```
let
Spoint = false -> dot   and pre S;
Splus  = false -> plus  and pre S;
S      = false -> varia and pre ( Uimpl or S or Splus );
U      = varia -> varia and pre ( Spoint or U or Uplus );
Uplus  = false -> plus  and pre U;
Uimpl  = false -> impl  and pre U;
tel
```

This program is presented here just to show the low complexity of the description.

4.2. Evaluation

In a similar way we have described the basic evaluator given by Jevons. It is quite obvious to describe it in Lustre. We have to repeat it one for each row of the truth table. Then the task of the Lustre compiler is to deal with a product-automaton obtained by product of the syntactic analysis automaton and the different basic evaluators.

4.3. Results

The global minimization of the product automaton gives 160 states for 2 variables and 5952 states for 3 variables instead of $4 * 4^8$ (3 variables yield 8 row truth table).

5. TOWARDS A VLSI IMPLEMENTATION

This machine [2] has been proposed to students as a project subject. The goal was to study an hardware implementation. It is of course convenient to deal with basic evaluators working in parallel. The direct synthesis of the global automaton is obviously a wrong idea.

The syntactic analyser "distributes" subject vs. subject' information to all the evaluators. The syntactic analyser is a very simple four states automaton.
It needs two flip-flops and a bunch of gates.
It may easily be synthesized by automated synthesis tools.

Then the evaluator for each row of the truth table computes its sequence of values by the aforementionned 4 state automaton. We name B0, .., B15 these evaluators.

This hardware is rather simple and can be done in a short time. An interesting part is to establish the "personalization" of each individual evaluator B0 to B15 to deal with the values of a'b'c'd', a'b'c'd, ...abcd rows. This is left to the reader as a lab activity.

6. CONCLUSION

Nowadays evaluation of logic formula is based on Binary Decision Diagrams. The programs based on this technique are very efficient from the point of view of the trade-off between space and time. Jevons' solution is obviously THE best solution for time. And the worst one for space! But when it was built, in 1870, the scale of integration was so large that the author could ignore this aspect !!! We must not forget that Jevons' machine with its 16 evaluators may be seen as the first Single Instruction Multiple Data machine (SIMD). Jevons himself had the idea of using truth tables.

ACKNOWLEDGEMENTS

I would like to acknowledge my colleague Nguyen-Huy Xuong, Professor of Discrete Mathematics, who helped me with regular grammars. David Alexander and Jacques Taillebois are two students working on a project dealing with Jevons' machine. Their questions and comments were very helpful. David was able to get a copy of the original text. Thanks to him. Jacques Cohen, Professor of Computer Science at Brandeis University, has provided helpful comments on the original manuscript.

REFERENCES

[1] P. Amblard, *Scheduling problems while compiling the real time language Lustre on the digital signal processor ST 18930*, Euromicro Workshop on Real-Time, Como (Italy), June 1989, pp 178-186

[2] G. Boole and W. Stanley Jevons, *Alg bre et Logique: D'apr s les textes originaux de G.B et W.S.J., avec les plans de la Machine Logique : Introduction et adaptation francaise par Frederic GILLOT*, Ed Albert Blanchard, Paris, 1962. (contains French translation of [7])

[3] Michel Breal, *Ethymologies grecques (Automatos) Mémoires de la Société de Linguistique de Paris*, Tome 10, 1897-1898, pp 402-403

[4] M. Gardner, *Logic Machines*, Scientific American, March 1952, pp 68-73

[5] M. Gardner, *Logic Machines and Diagrams*, The Harvester Press, Brighton, 1983, pp 91-103

[6] N. Halbwachs, P. Caspi, P. Raymond and D. Pilaud, *The synchronous Data Flow Programming Language LUSTRE*, Proceedings of the IEEE, Sep 1991, pp 1305-1320. Special issue on Real-Time Programming.

[7] W. Stanley Jevons, *On the Mechanical Performance of Logical Inference*, Philosophical Transactions of the Royal Society, 1870, pp 497-518

[8] L. Liard, *Les logiciens Anglais contemporains*, Ed. Germer Bailli re et Cie, Paris, 1878.

[9] W. Mays and D.P. Henry, *Jevons and Logic*, Mind, Vol LXII-1953, pp 484-505.

31

EDUCATIONAL COMPUTER USE IN THE MOS TRAINING FAB

[1]A. Ferreira-Noullet, [2]P.F. Calmon, [2]J.L. Noullet

[1]INSA - Institut des Sciences Appliquées de Toulouse
DGEI - Département de Génie Electrique et Informatique
[2]A.I.M.E. - Atelier Interuniversitaire de Micro-électronique de Toulouse
Campus INSA - Complexe Scientifique de Rangueil - 31077 Toulouse
FRANCE
Phone: (+33) 5 61 55 98 88, FAX: (+33) 5 61 55 98 00
E-mail: ana@aime.insa-tlse.fr

ABSTRACT

This work presents the interaction between the training on MOS fabrication in an educational clean-room at A.I.M.E. in Toulouse and the related software tools. The professional simulation tools are involved in demonstration sessions to show students the process condition parameters related to each process step. In addition, a home-made software is presented, which is used to manage the "Capacitance-versus-Voltage" measurements at the end of the circuit manufacturing. All these software tools are used in two contexts: to validate the process steps and to provide a guide to the students during the process handling.

1. INTRODUCTION

As mentioned in previous works [1,2,3], the clean-room at A.I.M.E. (Atelier Interuniversitaire de Micro-électronique) was created to provide to more than 400 students per year a practical training on basic MOS integrated circuit manufacture.

During the training period, the students are led to fabricate MOS integrated circuits from a blank wafer.

Because of the short training duration (one week), the process is kept very simple, only N-MOS based circuits are manufactured using a four-mask set, self-aligned polysilicon gates and two insulated levels of interconnect (aluminium and polysilicon).

In the wafer there are simple chips containing basic devices as MOS capacitors, MOS transistors, diodes and resistors, as well as more complex circuits with 10-micron channel length like a ring oscillator, a D flip-flop, a Schmidt Trigger and basic logic gates.

In parallel with the fabrication, the students characterize each process step and they must compare the measurement results with the simulations performed with the actual process parameters (for instance temperature, time, gas pressure and flow rate in the case of the oxidation process).

Meanwhile, during the breaks left by the clean-room activities themselves (waiting time for an oxidation, polysilicon deposition, etc.), the students attend a demonstration of the process simulation.

The demonstration is made at the CAD room at the A.I.M.E. using workstations and the SILVACO tools. The SILVACO tools include the SUPREM IV program for the process step simulations and the ATLAS program for the electrical device behavior simulation from the physical data obtained by SUPREM.

At the end of the training the electrical measurements on the final devices complete the information. Also, a big effort has been made in the C(V) ("Capacitance versus Voltage") characterization due to its relevance in the production quality monitoring.

The comparison between simulated and measured C(V) is made by the home-made software and the results are presented.

2. THE PROCESS STEPS

At the process beginning, each two-student groups receives three blank wafers, boron doped at 10^{15} atm/cm^3, <1 0 0> oriented. Only one of the wafers will contain the future manufactured chips, the other are used as process monitoring samples, so called "witness", that is, all the characterization steps will be made on these wafers.

The characterization starts with the wafer thickness measurement and initial doping by the four-probe method measurements.

After a deep cleaning step, the wafers are placed onto the oxidation furnace under the conditions presented on Table 1.

Table 1. The wet oxidation conditions

Temperature	Time(min)	Gas	Flowrate (liter/min)
Initial = 800°C, Final =1100°C rate = 12°C/minute	25	N_2	1.0
1100 °C	40	H_2 O_2	2.7 1.5
1100 °C	30	O_2	2.2
1100 °C	10	Ar	1.5
Initial = 1100°C, Final =800°C rate = -12°C/minute	60	N_2	1.0

At the clean-room, the students start to measure the visible oxide thickness in the "witness wafer". A 1cm slice is cut from the "witness" for this step. The slice is half etched chemically by use of the HF acid. The step produced is measured by means of the mechanical profilometer TENCOR and measured again by optical way using the elipsometer equipment.

The thickness values measured in the "witness wafer" usually match the simulation results, typically 500nm.

Thus, the students make themselves the photolithography step, the chemical etching of the field oxide and the R.C.A. cleaning. Afterwards, the dry oxidation is executed according to the new conditions shown on Table 2.

Table 2. The dry oxidation conditions

Temperature	Time (min)	Gas	Flow rate (liter/min)
1100 °C	30	O_2	2.0
1100 °C	10	Ar	2.0

The students do the same procedure of thickness characterization for the wet oxide. The measured values match the simulation results with a ±10% tolerance. A 70nm thickness is currently obtained.

The following steps include the polysilicon deposition by LPCVD (Low Pressure Chemical Vapor Deposition) methods. The polysilicon is phosphorus doped "in-situ" during the LPCVD process.

Thus after a photolithography step, the polysilicon layer is etched by RIE (Reactive Ion Etching) and the gate oxide is chemically etched.

One of the more important step of the MOS technology is the drain and source definitions. In our case, the junctions are made by the diffusion process.

The diffusion furnace follows the conditions on the Table 3.

Table 3. The diffusion conditions

Diffusion - Pre-Deposition Step			
Temperature	**Time (min)**	**Gas**	**Flow rate**
1050 °C	5	N_2	2.0 l/min
1050 °C	5	N_2 O_2 $POCl_3$	l/min l/min 5 mg/min
Diffusion - Redistribution Step			
1100 °C	10	N_2	1 l/min

We use an optical method to determine on a quick way a rough estimation of the PN junction depth. This method includes a wafer erosion in a particular shape (cylindrical erosion) and a coloration of the wafer surface with a developer which allows a color contrast between the N and P regions.

A more accurate measurement can be made by using the scanning electron microscope on a cross-section.

The simulated depth is located between 1.2 and 1.6µm as it can be read on the concentration profile (Figure 2).

The students' measurement range is between 1.6 and 2.2µm. Usually, the scanning electron microscope image shows a junction depth of 1.5µm. Then we can deduce that the mismatch is due to the visual observation of the NP junction.

The SILVACO software is used to display each step of the manufacturing sequence. This kind of representation helps students to understand the process.

Figure 1 is the last image of the collection, showing the cross section of a finished MOS device (note that the horizontal and vertical scale are different).

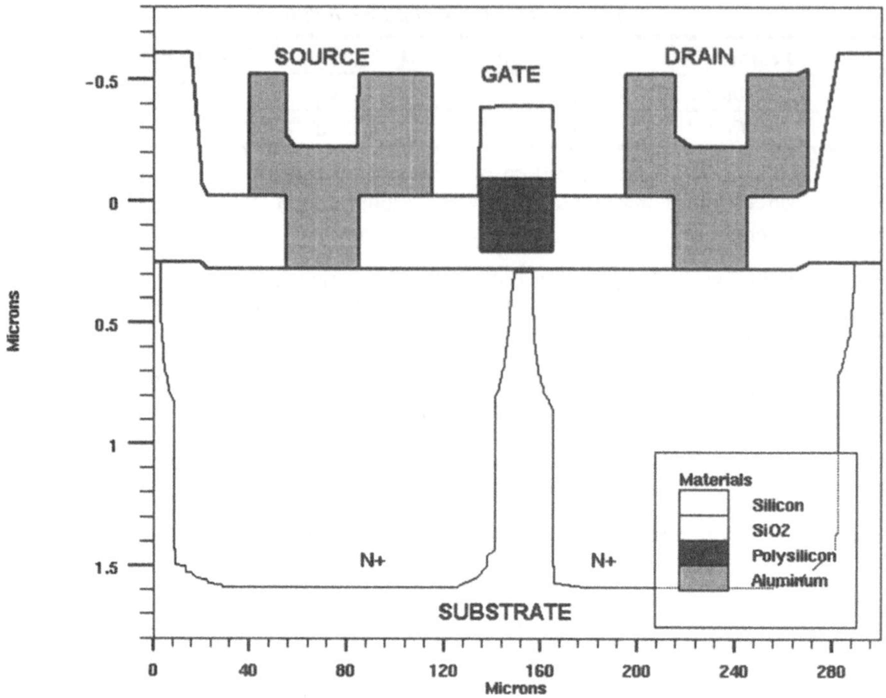

Figure 1. The simulated view of the finished MOS device

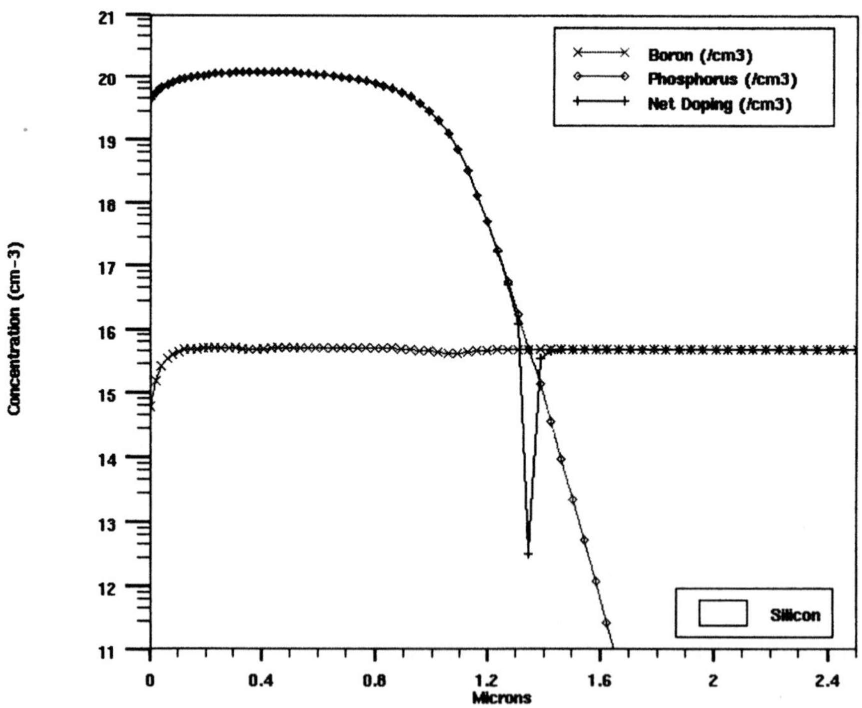

Figure 2. The N-P junction concentration profile simulated by SUPREM IV

3. THE CAPACITANCE VERSUS VOLTAGE MEASUREMENTS

The measurement **Capacitance = f(voltage)** in the MOS structures, named **C(V)**, allows to check the quality of some critical technological steps necessary to the MOS manufacture. These steps are the surface cleaning in the oxide-semiconductor interface, the thermal oxide growth and the gate material deposition procedures.

At the end of each training, the students perform a computer-aided **C(V)** measurement on a characterization chip cut from their wafer, in order to check the technological parameters.

The **C(V)** experimental curve is compared to the theoretical one computed from given data like:

- the doping type and concentration of the silicon substrate,
- the permittivity and thickness of the insulating material

Furthermore, the semi-automated fitting of the C(V) curve allows the extraction of the following parameters:

- the gate material surface and type.
- corrected substrate doping (from measured C_{max} / C_{min})
- corrected MIS structure area (from insulator thickness and measured C_{max})
- flat band capacitance (from doping and area)
- flat band voltage (from flat band capacitance)
- threshold voltage (from theoretical threshold and flat band voltage)
- fixed charge density (from flat band voltage)
- mobile charge density (from voltage hysteresis)

The software features a built-in support for parasitic capacitance correction, based on the measurement of two capacitors on the same chip, with different geometries.

There is also a support for stress test. The stress test, or accelerated aging, is the sequence of:

- initial C(V) measurement and results storage on disk
- some 20 minutes at high temperature (200°C) and near breakdown voltage (20V)
- new C(V) measurement, and enhanced extraction of mobile charge density.

The C(V) test bench is made of the HP4192A instrument connected to a PC computer running the home made software developed by means of the LABWINDOWS-CVI toolkit. This software offers a user-friendly graphical interface with mouse support and built-in help (see Figure 3).

In addition, there is a "virtual lab" version of the software that can be installed on any PC, without the HP instrument, for training purpose. In this version, the physical measurement is emulated using previously stored measurement data.

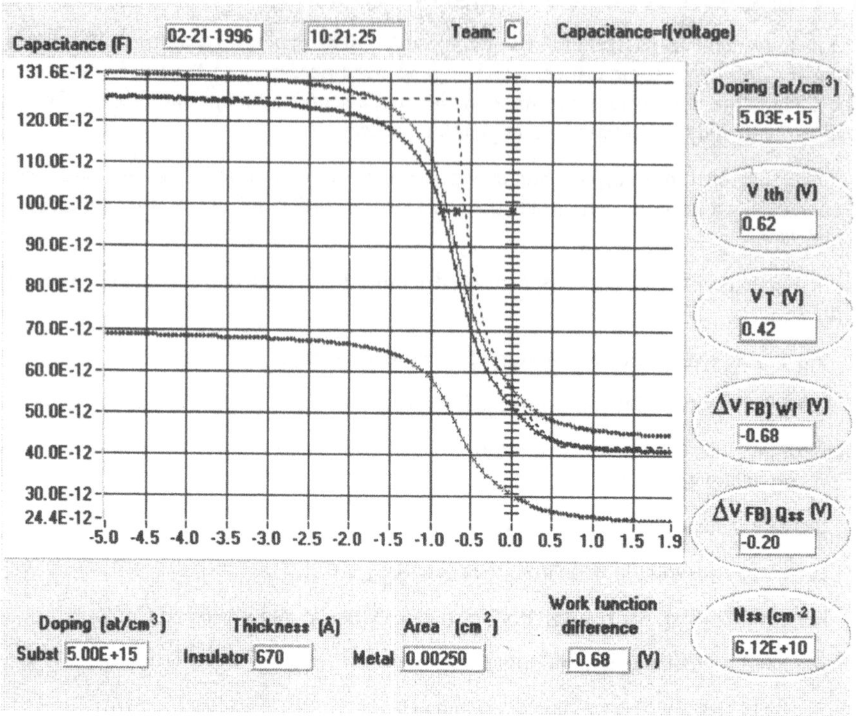

Figure 3. The C-V test using the home-made software

REFERENCES

[1] P.F. Calmon and G. Pierrel, *Caractérisation et test de composants. Automatisation de banc de mesures partir du logiciel Labview*, Troisi mes Journées Pédagogiques du CNFM, Saint Malo, France, Dec. 1994.

[2] J.L. Noullet, P.F. Calmon et al., *Introducing Digital Gates in the Basic MOS Fabrication Training*, 1st European Workshop on Microelectronics Education, Grenoble, France, Feb. 1996.

[3] J.L. Noullet, P.F. Calmon, A. Ferreira and G. Pierrel, *The Chips Manufactured by the Students: News from the MOS training Fab*, 3rd Advanced Training Course Mixed Design of Integrated Circuits and Systems, Lodz, Poland, May-Juin 1996

[4] P.Nouet, P. Lepinay and A. Ferreira, *Simulation de technologie: outils et utilisation pédagogique*, Quatri mes Journées Pédagogiques du CNFM, Saint Malo, France, Dec. 1996

32

TEACHING OF ANALOG IC DESIGN WITH MODERN CAD TOOLS AND CMOS PROCESSES

Jürgen Frickel, Jafaar Mejri and Wolfram Glauert

Institute for Computer Aided Circuit Design
University of Erlangen-Nuremberg
Cauerstrasse 6, 91058 Erlangen
GERMANY
phone: ++49 9131 8586-95, fax: -99
email: fric@lrs.e-technik.uni-erlangen.de
URL: http://www.e-technik.uni-erlangen.de/Home.html

ABSTRACT

Topic of this article is a course "Design of Integrated Circuits", which is provided for 10 students. The students are requested to develop, optimize and design a digitally controlled amplifier using switched capacitor techniques. The main target of this course is to familiarize the students with a modern design environment including modern CMOS processes and "state of the art" design tools (i.e. HSPICE, Cadence DFW II). We try to provide the students with practical experience in mixed-signal design and develop their capabilities to solve design problems.

1. MOTIVATION

Electrical engineering students at the University of Erlangen-Nuremberg are requested to attend three courses of laboratory work during their advanced studies. Our institute, researching and teaching in the domain of IC design and test, offers three courses for students, which are:

- Design of Integrated Circuits [6]
- Modelling and Synthesis of Integrated Circuits [3] [4]
- Testing of Integrated Circuits

The topic of this article is the course "Design of Integrated Circuits". The main goal of this course is to introduce the students into the design of mixed-signal integrated circuits on circuit and layout level. Another goal is to familiarize the students with "state of the art" design tools such as Cadence DFW II or HSpice (from EUROPRACTICE initiative) and to provide some design experience with modern industrial CMOS processes.

Starting with a loose description of the function of the circuit the students have to work out a detailed specification and topology for the involved circuit modules. The modules are independently simulated, optimized and physically designed (layout). The modules are then combined into a complete design, which will be checked and verified using simulation and verification tools.

The purpose of the course is to provide the students with practical design experience and to develop their capabilities at solving design problems such as:

- Analysis of a technical problem and specification of a design
- Discussion of alternative implementations and circuit topologies
- Modelling and system simulation of the different topologies
- Optimization and verification of modules by means of circuit simulation
- Floorplanning: Integration of designed blocks into the complete system
- System verification
- Project work, team work, communication between working teams

2. COURSE ENVIRONMENT

The course has a capacity of 10 students and lasts 3 months. A weekly attendance of 5 hours was required, the main discussion sessions and introductions took place during the scheduled presence time. The students had discretionary access to available workstations during the 3 month period.

The students were split in groups of two, every group was requested to deliver a complete full custom ASIC design at the end of the course.

As a prerequisite to the course and in order to achieve acceptable results the students had to attend a lecture of 1 semester dealing with the design of integrated circuits. The students attending the course were in their sixth or eighth semester.

3. DESIGN OF A VARIABLE-GAIN SWITCHED-CAPACITOR AMPLIFIER

The starting information provided for the students was the task to build a digitally controlled variable-gain amplifier for audio frequencies. One restriction was the use of a given CMOS technology.

A parallel lecture held during the design course introduced the students into the design of CMOS operational amplifiers and their different uses. The concepts of switched capacitors and their ability to emulate a resistive behaviour were also discussed during the lecture (see Figure 1, week 1-3).

The students had to discuss the implementation of the variable gain amplifier using feedback resistors and the inherent drawbacks of this circuit configuration. The lack of accuracy and the prohibitive layout area needed for the circuit led to the use of switched capacitors in order to emulate the resistors needed in the circuit.

After having reached an agreement on the topology of the amplifier implementation, the students had to specify the needed sample frequency (imposed by the frequency range of the circuit) and the feedback topology of the amplifier (imposed by the specified 3 dB point of the amplifier). A digital control scheme was defined and the digital switch control was designed. The next step was a preliminary design of a switched capacitor unity resistance.

After an introduction to the HSpice circuit simulator, the students were able to simulate their ideas of the designed circuit (see Figure 1, week 4).

The simulation and analysis of the behaviour of the switched capacitor led the students to a different structure, which is less sensitive to parasitic capacitances [1]. The next step was the design of a control circuitry for the SC-switches at the transistor level, which provided control signals for the transmission gates and accordingly delayed signals for the switch controls. Next came the refinement of the switch transistors, which should allow a fast loading of the unity capacitance used in the SC-"resistor". Finally for this part the simulation of the whole SC-"resistor" module together with the control circuitry followed.

The specification for the operational amplifier was agreed upon after having simulated the amplifier circuit using a behavioural model and taking account of the fact that the resistances consisted of SC-"resistors" (e.g. slew rate imposed by the capacitance at output of the amplifier). The operational amplifier was then designed using the above mentioned specifications [2].

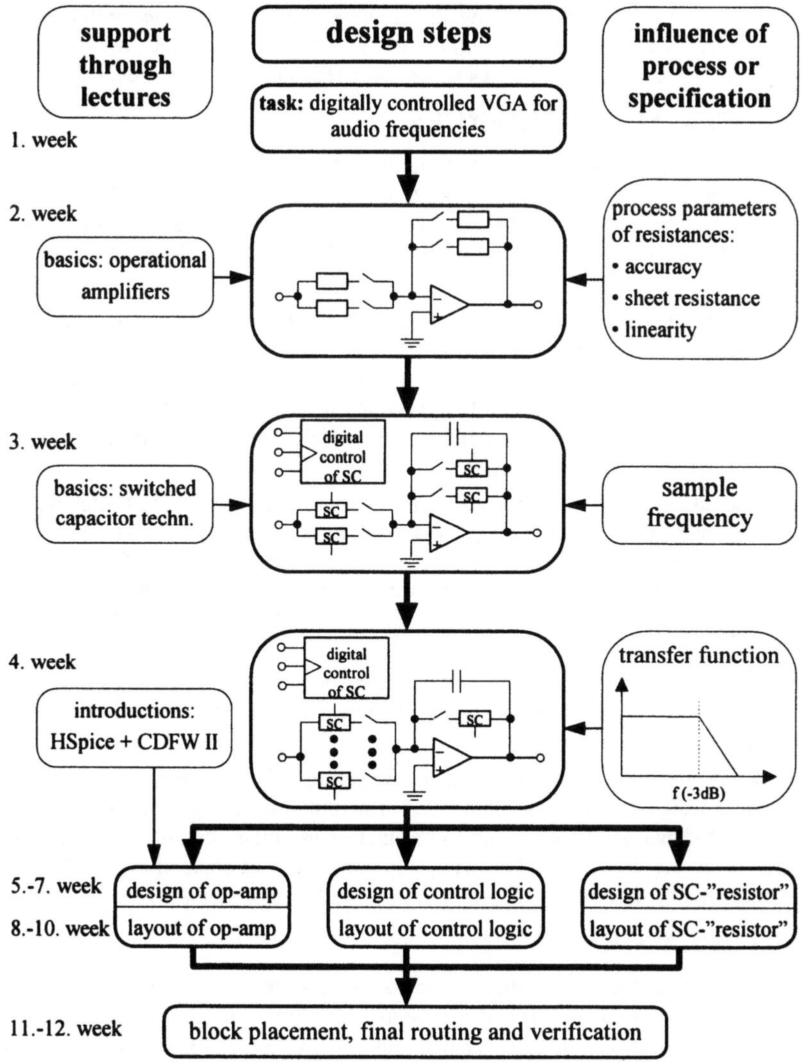

Figure 1. Lectures, time and flow plan of the circuit development

After an introductory session with Cadence DFW II, the students were able to enter the schematics of the designed circuits and to start the layout phase. The students were requested to layout the modules independently and to couple together only layout blocks with a verified functionality. Therefore the students had to make a full extraction of the layout and had to run simulations for verifying the functionality of the layout. A layout-versus-schematics (LVS) of the extracted circuit had also to be carried through as a final verification step.

The students designed independently the layout of the different modules (see Figure 1, week 8-10), but had to take care of the communication flow within the team. Preceding the layout entry phase a discussion of the layout planning and layout strategy had to take place. During this session the students had to specify the partitioning and placement of the different modules in order to minimize the parasitic disturbances.

The final work consisted of gathering the layouted modules into one circuit layout and performing the remaining signal routing. Finally the whole circuit was extracted and a final layout-versus-schematics check as well as a HSpice simulation closed the course.

4. PRACTICAL EXPERIENCES

The course has been held three times at the Institute for Computer Aided Circuit Design. It has been observed that the participating students attained a deeper knowledge in the area of integrated circuit design [3] [4], as the grades achieved in the examinations in the corresponding lectures were noticeably better than those of the students only attending the lectures.

The course led the students to a better understanding of analog circuits and gave insights into the design of standard cells. All the students were able to cope with the complex design rules of a modern CMOS process. The students were able to go through all the design phases from the specification to the actual layout of the circuit. The groups needed a weekly average of 10-14 hours to complete the circuit. This was in part due to the longer simulation time needed for SC-circuits.

Since the additional work the students performed was not supervised, it allowed them to arrange their schedules (e.g. attending lectures) in a more flexible way. The completion of such a complex circuit was mainly dependent on the wide availability of the workstations for the students. The fact that the students were allowed to work on their designs without time constraints guided to a more uniform distribution of the staff's work load over the week.

The supervising staff had to insure that the students worked symmetrically in the group, as the students tended to take "roles" e.g. the "layouter" and the "designer" in the final phase of the circuit design.

The loose specification of the circuit at the beginning of the course allowed the staff to show typical design pitfalls to the students. Furthermore in having to specify the complete circuit, the students developed much more interest in "their" circuit which led to more engagement during the design. The winner in some kind of competition on the smallest layout between the different design groups reached an area of about 0,30 x 0,29mm.

5. FUTURE WORK

In the future different mixed signal circuits are planned for the course. The use of VHDL-AMS (analog mixed signal hardware description language) is also planned, a VHDL-AMS course will then be offered within an interactive hypertext teaching system [5]. Other tools and simulators (for VHDL-AMS) will be used as soon as they are available at the Institute for Computer Aided Circuit Design.

REFERENCES

[1] R.L. Geiger, P.E. Allen and N.R. Strader, *VLSI Design Techniques for Analog and Digital Circuits*, McGraw-Hill, 1990, ISBN: 0-07-100728-8

[2] P.R. Gray and R.G. Meyer, *Analysis and Design of Analog Integrated Circuits*, John Wiley & Sons, 1993, ISBN: 0-471-87493-0

[3] J. Frickel, U. Heinkel, C. Kuntzsch and M. Selz, *Design of a Transmission Failure Indicator Chip in a Student Course Using VHDL*, 5th Eurochip Workshop on VLSI Training, Dresden, Germany, 1994

[4] J. Frickel and U. Heinkel, *TRAFFIC - Designing Chips for Digital Transmission Tests in a Student Project*, Education of Computer Aided Design of Modern VLSI Circuits, MixVLSI, Kraków, Poland, 1995

[5] J. Frickel, U. Heinkel, M. Padeffke and W.H. Glauert, *A Hypertext-Based Interactive Teaching System for Designing Integrated Circuits with VHDL*, 3rd East-West Congress on Engineering Education, Gdynia, Poland, 1996

[6] J. Frickel, J. Mejri and W.H. Glauert, *Introduction of Students to Modern CAD Tools and Processes for Analog CMOS Design*, Mixed Design of Integrated Circuits and Systems - Education of Computer Aided Design of Modern Devices and ICs, Poznan, Poland, 1997

33

TEACHING POWER ELECTRONICS WITH TWO-DIMENSIONAL SEMICONDUCTOR DEVICES MODELS

Mariusz Grecki, Grzegorz Jabłoński, Marek Turowski and Andrzej Napieralski

Technical University of Łódź
Department of Microelectronics and Computer Science
Al. Politechniki 11, 93-590 Łódź
POLAND
e-mail: mops@dmcs.p.lodz.pl

ABSTRACT

This paper presents possibilities of application of multidimensional physical simulation in semiconductor devices teaching. The simulation examples obtained using simulation program MOPS (Modelling Program for Semiconductor Devices) are presented. They show the phenomena which can not be presented to students using SPICE-like circuit simulators. The MOPS program is used in teaching process in the Technical University of Łódź.

1. BEHAVIOURAL VS PHYSICAL MODELLING

The modern technology of electronic circuits manufacturing requires from the designers deep understanding of semiconductor devices operation principles. This knowledge can not be gained from utilization of typical programs for analysis of electronic circuits like the most popular SPICE. Such programs use lumped models of semiconductor devices and allow to obtain only voltages and currents in the circuit. This approach does not allow to look "inside" the device to understand the physical phenomena. The only way to investigate them and present to the students are numerical simulations using physical models of semiconductor devices. However, they are hard to perform during typical laboratory exercises. The model of carriers behaviour in the semiconductor consist of complex spatial differential equations. They are non-linear because of the non-linear dependency of physical parameters (such as carriers mobility, generation-recombination intensity, carriers life-time) on the carriers density, electric field and temperature [1]. The solution of this set of equations is not easy to obtain and requires advanced numerical methods. The physical simulation of semiconductor devices is a very time consuming task. It is caused by solution of large set of non-linear partial differential equations. This problem is particularly serious for multidimensional analysis. Apart from strictly computational problems there are two factors especially important in teaching process. First, the programs must be easy to use for inexperienced users, students in particular. Second, they should provide clear visualization of simulation results.

There exist many programs for physical simulation of semiconductor devices, like BAMBI, MINIMOS, PISCES, MEDICI, but their features make them not applicable for our purpose. They are expensive or require at least workstation class computer to run. Additionally, they expect sufficiently prepared user. Therefore there is a need to develop specialized educational physical simulation programs.

A few years ago existing personal computer capabilities allowed to perform only one-dimensional simulation. For a limited set of semiconductor devices - diodes and thyristors - multidimensional effects can be neglected in case of qualitative analysis sufficient for educational purpose. Unfortunately, there is still wide range of devices which cannot be analyzed this way. In case of field-effect devices, horizontal field generated by gate influences conductive properties in vertical direction. The analysis of such devices requires to use at least two-dimensional models. This kind of analysis is also necessary to show specific problems appearing in planar devices (e.g. in integrated circuits). Moreover, it enables to present two-dimensional effects influencing the behaviour of power devices (e.g. current squeezing).

Fortunately, the rapid growth of personal computer performance makes possible to develop a simulation program for two-dimensional analysis of semiconductor devices to be run on a PC. With the help of such a software tool we can demonstrate principle of operation and internal behaviour of almost all semiconductor devices.

The MOPS (**Mo**delling **P**rogram for **S**emiconductor Devices) developed at our institution, is aimed at analysis of electronic circuits containing semiconductor devices modelled by two-dimensional physical models, based on simulation of phenomena inside the semiconductor structure. Program can be run on IBM-PC class computers under Linux operating system or Windows NT. Special versions of this program, without the possibility of graphical display, may be run on any Unix or MS-DOS system. The MOPS program solves set of transport equations, including Poisson equation and continuity equations for electrons and holes [2]. The solution is obtained numerically after discretization of the equations on the automatically generated *finite-boxes* mesh [3].

The program offers a very simple user interface using X-Windows system. Figure 1. presents a screen snapshot of MOPS program during the simulation of an IGBT. In the off-state in the wide base area of IGBT the density of holes is low because this region is lightly n-type doped (Figure 1a). On the other hand, there are no electrons in the channel area (because of low gate voltage and no accumulation - Figure 1b). In effect, the device prevents current from flowing. This kind of phenomena cannot be shown using lumped models of semiconductor devices (used e.g. in SPICE program).

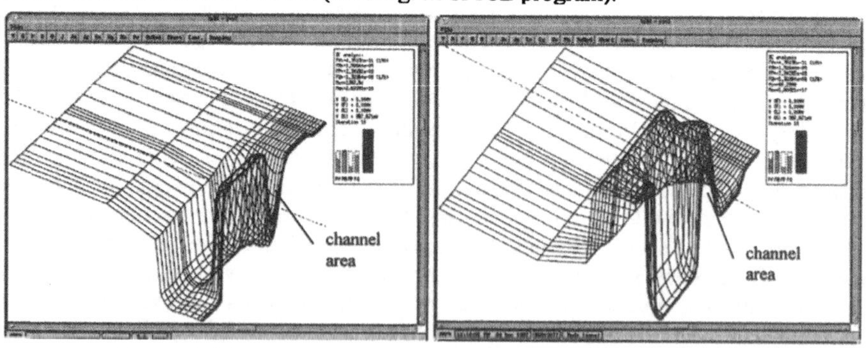

Figure 1. The carriers density in the IGBT structure in OFF-state
a) holes *b) electrons*

2. SIMULATION A OF BIPOLAR TRANSISTOR

As an educational example, we will show the results of analysis of switching process of power bipolar transistor, proving that lumped models of such devices are very inaccurate. Therefore, multidimensional physical models should be used in this case.

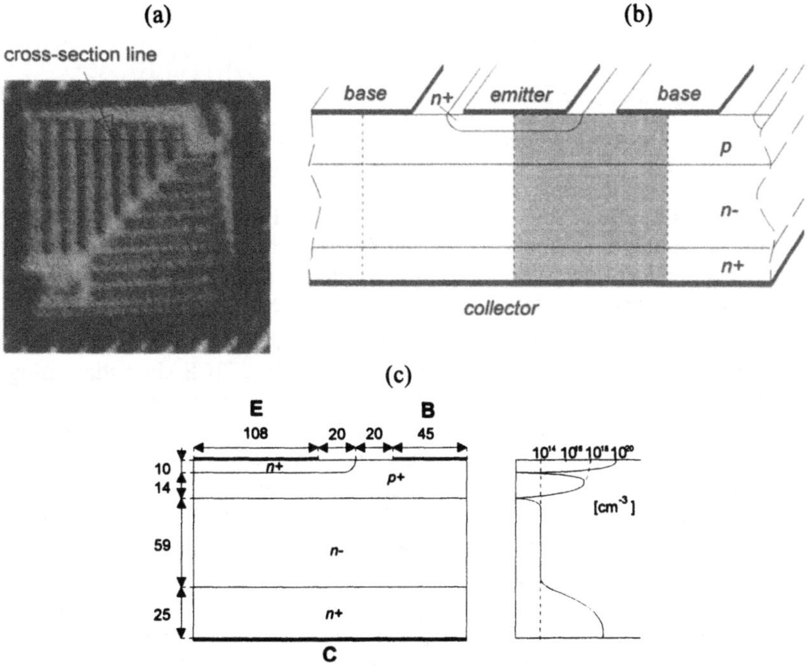

Figure 2 Transistor BUX 48:
a) structure microphotograph
b) cross-section of the structure (grey area = elementary cell)
c) cross-section and doping profile of elementary cell of the transistor
(all dimensions in μm)

A two-dimensional model of the transistor has been created using technological data of BUX 48 transistor. The cross-section and doping profile of elementary cell of the analyzed transistor are presented in Figure 2. The simulated circuit consists of the transistor switching current in a 25Ω resistive type load (with very small 80nH parasitic inductance of wires) forced by 300 V voltage source. The results of simulation are very interesting. It has occurred that the current distributions in steady-state and transient conditions differ significantly [4].

In Figure 3 the current distribution in the transistor structure is presented, single strip corresponds to 10% of total current. Under DC conditions the current distribution is fairly uniform and we can reduce the model of the transistor to only one dimension. During fast switching, the current flow is not uniform because of the limited speed of carrier diffusion. During turn-on, the current flows mainly through the path near the base contact as in this area the conduction is initiated by carrier injection. During turn-off, a reverse situation occurs, and the current flows through the path placed under the centre of the emitter contact. In this area the carriers are removed just at the end of the turn-off process, in contrast to the region near the base contact where it is much easier to remove the carriers by base current.

(a) (b) (c)

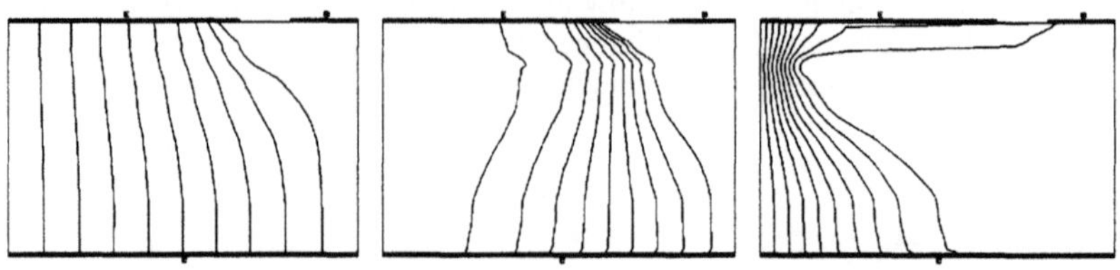

Figure 3 Current distribution inside the transistor structure during DC (a), turn-on (b) and turn-off (c)

The above considerations show that current flow inside the transistor structure during switching has a multidimensional nature and for transient processes its model cannot be reduced to one dimension without loss of correctness. It means that even a distributed one-dimensional physical model is inadequate in this case.

In order to show the inaccuracy of lumped SPICE-like models we will compare the results obtained from two-dimensional simulation (Figure 4), the SPICE simulation using BUX 48 model from *euro.lib* PSPICE library (Figure 5) and the experimental data (Figure 6). The parameters of switching process are summarized in Table 1. It can be clearly noticed that all characteristic time parameters (delay, rise, storage and falling time) foreseen by two-dimensional simulation are close (within 25% accuracy) to the measured ones. These errors result from the inaccuracies in spatial distributions of doping (for the simulations the idealized Gaussian model was used) and other physical parameters. The results obtained from SPICE simulation differ significantly, particularly rise and falling times (130% error). The situation is even worse if we compare the power losses. SPICE predicts the total energy dissipated during switching process almost 10 times higher than two-dimensional simulation. This is due to unrealistically high collector-emitter voltage as shown in Figure 5. The waveforms obtained from two-dimensional simulation agree well with the result of measurements as illustrated in Figure 4 and 6.

(a) (b)

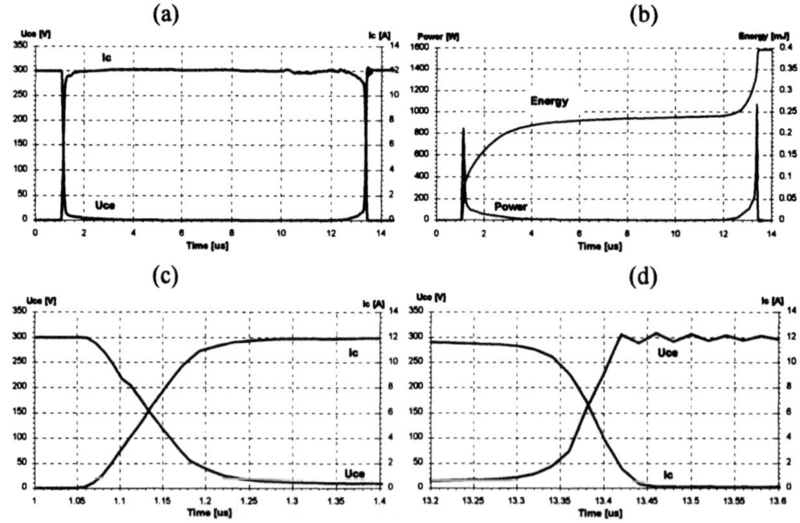

Figure 4 BUX 48 switching - 2D simulation results
a, b) entire switching process, c) turn-on, d) turn-off

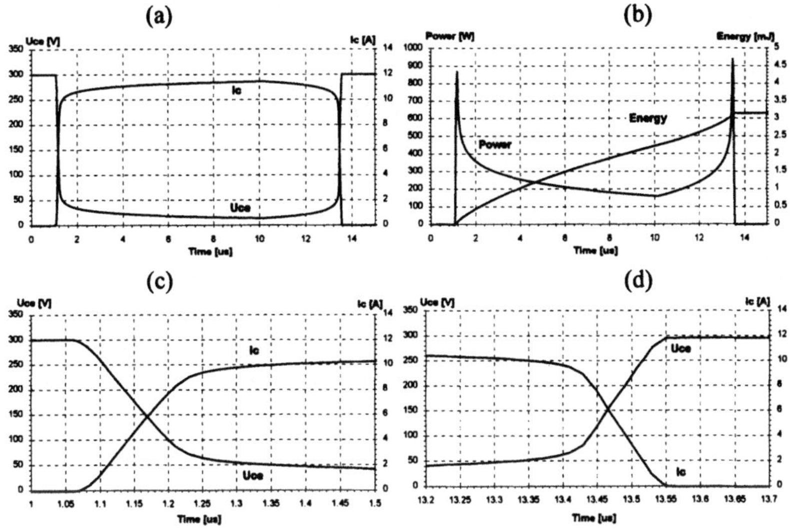

Figure 5 BUX 48 switching - SPICE simulation results
a, b) entire switching process, c) turn-on, d) turn-off

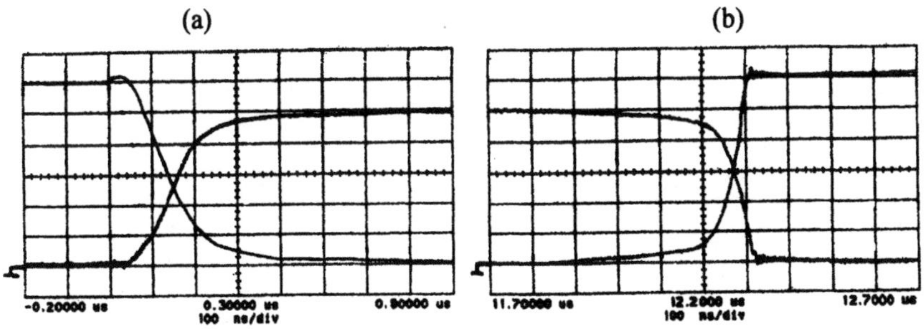

Figure 6 BUX 48 switching - experimental results
(collector current scale 2A/div, collector-emitter voltage scaling 50 V/div)
a) turn-on, b) turn-off

Table 1. BUX 48 switching process parameters

	t_d [ns]	t_r [ns]	t_s [μs]	t_f [ns]
2D Simulation	80	132	3.3	91
PSPICE Simulation	105	405	3.28	247
Experimental data	70	175	3.2	110

3. CONCLUSION

A two-dimensional simulation program has been used by students in our institution during the course "Modelling of Integrated Circuits". Carrier distribution and current flow in MOSFET's and BJT's have been observed. We believe that two-dimensional simulation significantly improves understanding of physical processes occurring inside semiconductor devices, enhancing process of semiconductor physics education. The ease of use and friendly interface is a necessary condition of efficient educational application of a program. The current work is aimed towards adapting our existing 2D semiconductor simulation tools to be applied in the teaching process in the most effective way.

REFERENCES

[1] Dorkel J.M. and Leturcq P., *Carrier mobilities in silicon semi-empirically related to temperature, doping and injection level*, Solid-St. Electron.,vol. 24, 1981, s.821-825

[2] Turowski M., *Two-dimensional simulation of power semiconductor devices*, PhD. Thesis, Institute of Electronics, Technical University of Łódź, Łódź, 1992

[3] Franz F.F., Franz G.A., Selberherr S., Ringhoffer C. and Markowich P., *Finite Boxes - a generalization of the Finite Difference method suitable for semiconductor device simulation*, IEEE Trans. Electron Devices, vol. ED-30, No. 9, 1983, s.1070-1082

[4] Napieralski A. and Grecki M., *Optimal design of power bipolar transistor*, Proc. 5th European Conference on Power Electronics and Applications, Brighton (UK) 1993, vol.2, s.396-401

[5] Napieralski A. and Napieralska M., *Polowe półprzewodnikowe przyrządy dużej mocy*, WNT 1996, Warszawa (in Polish)

34

VLSI TOP-DOWN DESIGN FOR STUDENTS OF COMPUTER SCIENCE - A PRACTICAL COURSE

Igor Katchan, Frank Mayer and Detlef Schmid

University of Karlsruhe
Institute of Computer Design and Fault Tolerance
P.O. Box 6980, D-76128 Karlsruhe
GERMANY

ABSTRACT

In this section a practical course on advanced methods of top-down hardware design is presented. The course is offered at the Computer Science Department at the University of Karlsruhe. It focuses on hardware modelling with VHDL and synthesis with the professional CAD tool set of Synopsys [1]. The topics of several practical lessons cover various hardware applications including neural networks and test controllers.

1. INTRODUCTION

Years of intensive academic and industrial research have resulted in various new methods for automatic hardware design. These include hardware description languages as well as numerous synthesis algorithms for the transistor and logic level up to the register transfer and behavioural level [2, 3], which have been merged and implemented in industrial CAD systems. Also, validation techniques have been improved and become more precise and, thus, more reliable. Hence, with today's fast computers and the actual CAD systems it is possible to correctly design hardware in short time.

Another important trend is the shift from the implementation oriented tasks of the more concrete lower levels up to the abstract higher levels with stress on the functional behaviour of a circuit. Therefore, the use of CAD tools is no longer limited to engineers with expert knowledge on electronics.

The behaviour of a circuit is usually specified by means of a hardware description language (HDL). One of the most popular HDLs is VHDL (Very high speed integrated circuits HDL), which is an IEEE standard since 1987. VHDL is "intended for use in all phases of the creation of electronic systems" [4]. Besides the functional and structural descriptions, it also supports various models of the timing behaviour of a circuit. The similarity of VHDL to well known imperative languages like Pascal or C makes it easy to learn for people who have some experience in programming.

Algorithms, functional behaviour, and programming are all topics of computer science. Hence, in 1993, we decided to develop a practical course on the basis of VHDL and a modern CAD tool, the Synopsys system, addressed in particular to students of computer science. It has been offered at the Department of Computer Science at the University of Karlsruhe since 1994.

In the following sections we first give some reasons why we chose the Synopsys tools, then the organization of the course is discussed followed by a description of the particular tasks and their educational intentions. At last, we summarize some experience we have got in five courses so far.

2. DESIGN TOOLS

An important problem of the preparation phase was the choice of a CAD tool for the course. The tool should be easy to use, run dependably, cover a wide range of design levels - especially the higher levels - support VHDL, and, last but not least, it should be inexpensive.

Non commercial tools are cheap, public domain programs are even for free. But their handling is difficult when, as usual, there is no graphical surface and the operating styles of tools for different design steps are not consistent. Usually, their maintenance is also poor, and reliable behaviour, especially when used by non experts, is critical.

Commercial tools are expensive. But they usually have clear graphical surfaces which are useful for several design steps. Therefore, they are easy to operate with. They are also well maintained and quite stable. Another advantage is the wide area of design levels they cover. Hence, they match our requirements for a practical course much better than public domain tools, and we decided to choose a commercial CAD system.

Among the commercial tools, the Cadence [5] and Synopsys tools were our favourites, because they were both supported by the European initiative Eurochip (currently Europractice [6]), which offers favourable conditions for using CAD software for educational and research purposes. Both systems are well known in the industry. The Cadence tools cover the whole range of hardware design steps from the behavioural level down to the layout. Additionally, it was already used at our institute in a practical course on traditional VLSI design. The Synopsys system ends at the gate level, but it is much more comfortable at the higher levels. As our practical course emphasizes the higher levels, we elected Synopsys and decided to use the Cadence tools only for the low-level design.

3. ORGANIZATION OF THE COURSE

The aim of this course is to give an introduction to the top-down design methodology, VHDL, and a modern CAD tool. Main subjects are the functional VHDL description of the circuit's behaviour based on a given verbal or formal specification and the validation by simulation.

The course is divided into 12 practical sessions, held once a week. Each of them lasts four hours. During the sessions, the students can use the computers (Sparc 10) in our lab and are supervised by a tutor. They have to cope with several tasks, each discussing another topic of hardware application and its referred problems. Together, the tasks cover the whole range from the behavioural level down to the layout. Preparation of the theoretical parts and the development of the VHDL programs have to be done in homework. The computer sessions are provided for editing, simulation, debugging, synthesis, and so on.

Our course is addressed, but not limited to graduate students of computer science. The students are supposed to have basic knowledge of imperative programming languages like C or Pascal, UNIX, and how to use a text editor.

4. DESIGN TASKS AND TOPICS

Before the students can start with any design task, they first have to learn VHDL. For this purpose we wrote a manual, "Introduction to VHDL", which is given to each student at the beginning of the course. The introduction is a short booklet where the syntax and semantics of those VHDL constructs are described which are necessary to solve the design tasks of the course. It also includes some hints for effective VHDL programming and examples of descriptions of combinational and sequential circuits. For further details on VHDL we recommend some comprehensive books like [7-9].

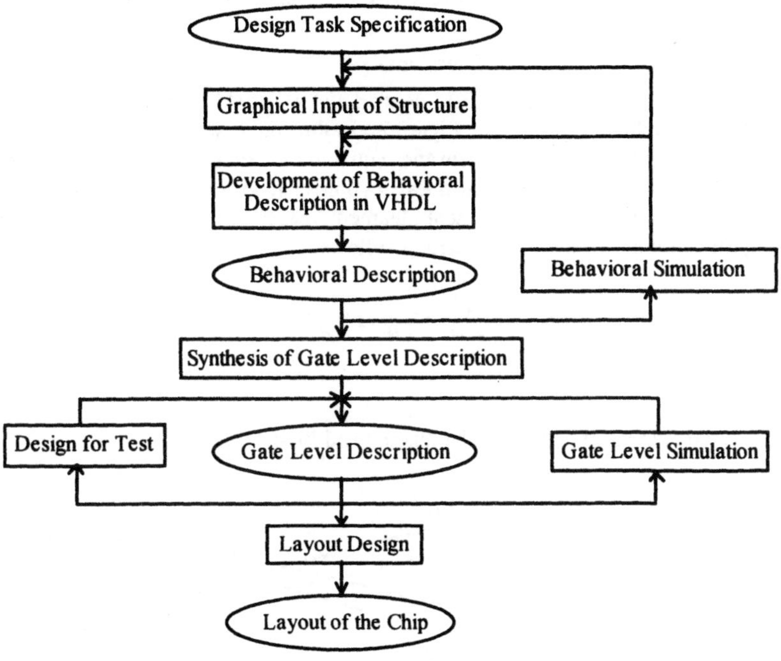

Figure 1. Design Process

The general solution scheme of the design tasks is shown in Figure 1. After studying the theoretical basics of a task and preparing some exercises, the students start with the graphical input of a given structure of the circuit they have to design. This is done within the Synopsys Graphical Environment (SGE). Basing on this graphical input, SGE automatically generates the VHDL code of the structure. The students then have to complete the structural VHDL frame by the behavioural description of the circuit. In the next step, they have to debug and test the code using a simulator. Depending on the design task, the synthesis of the circuit within Synopsys' "Design Analyzer" - in one task considering design for test - and the simulation of the resulting gate level description follows. Eventually, layout design is performed by the Cadence system.

The design tasks are as follows:

- Tutorial on Synopsys (1 session).

The course starts with a tutorial on the Synopsys tools. A detailed booklet guides the students through all design steps within the system which are part of the course. With the help of a small circuit example, the students become familiarized with the Synopsys tools and are confronted with all essential problems of VHDL programming they will come across in the following design tasks. The tutorial also highlights the VHDL synthesis subset of Synopsys. Like the "Introduction to VHDL", the tutorial is intended to be a reference work for the following design tasks.

- Pseudorandom pattern generator (1 session).

A simple pseudorandom pattern generator based on a linear feedback shift register (LFSR) is the first circuit the students have to design on their own. The functional behaviour is specified by an abstract matrix formulation. At the beginning, the students do not know that this formulation is nothing but a LFSR. The aim is to show the influence of a particular description style of the behaviour on computation expense and the quality of the synthesis results. The example also deals with sequential circuits and asynchronous reset signals.

- Hopfield neural network [10] (5 sessions).

The main topic of this task are Hopfield networks. A Hopfield network is a neural network with only one level of neurones (see Figure 2). In the course we consider a special kind of these networks with the number of neurones equal to the number of inputs as well as to the number of outputs. Each of the neurones is connected to all inputs. These connections have individual weights which represent „learned" pieces of information. Their values for learning (m-1)-bit patterns are calculated as follows:

$$w_{ij} = \sum_{s=1}^{p} In_{si} \cdot In_{sj}; \quad i \neq j$$

$$w_{ii} = 0,$$

where p is the number of patterns to be learned and $In_j \in \{-1,1\}$.

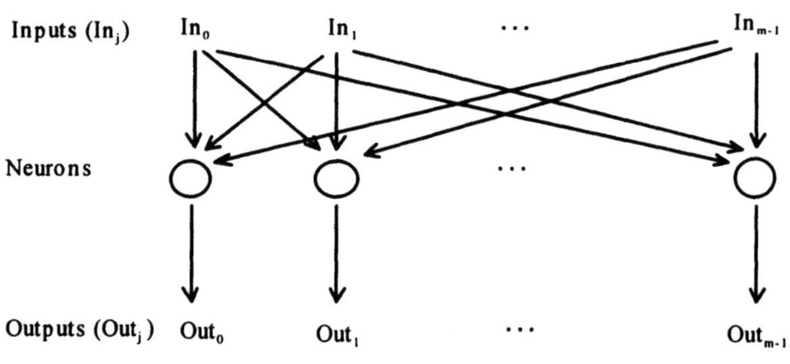

Figure 2. Hopfield Network

The network is able to recognize previously „learned" patterns even if they have been partly distorted. In the recognize mode the network output values Out_j at time t are determined by:

$$Out_i(t+1) = sign \sum_{j=0}^{m-1} w_{ij} \cdot x_j; \qquad x_j = \begin{cases} In_j; & t=0 \\ Out_j(t); & t>0 \end{cases} \qquad sign(y) = \begin{cases} 1; & y \geq 0 \\ -1; & y<0 \end{cases}$$

The pattern is considered as recognized if $Out_j(t+1) = x_j$ for all j.

The students have to develop and implement the learning and recognizing algorithms of the network as well as an input-output interface. The internal weights of the network are assumed to be stored in an external memory. The network circuit should be able to communicate directly with the external memory as well as via a bus arbiter. Memories may have an access time of one or more cycles. Handshake mechanisms, race conditions, hazards on signals, and three-state bus drivers are additional subjects of this task.

- Boundary-Scan Technique [11] (4 sessions).

Design for test tools are an obligatory part of modern CAD systems. Various kinds of scan path techniques are the most important of them. In this task, the students first have to integrate Boundary-Scan circuitry into a previously designed circuit (1 session). Their second task is to design and validate a Boundary-Scan Test Controller (see Figure 3). Additional topics of this design task are serial/parallel and parallel/serial converters as well as multiple clock signals which are sensitive to the rising and falling edges.

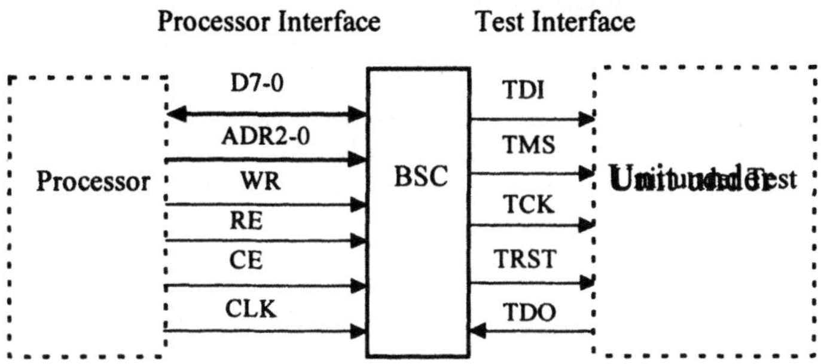

Figure 3. Boundary-Scan Controller

- Layout Design (1 session).

The course finishes with the layout synthesis of a formerly designed chip. The students are introduced to the CAD system Cadence, again with the help of a tutorial. This task is just intended to complete the whole design process.

5. CONCLUSIONS

We described a practical course on VHDL and the Synopsys system, which has been offered - meanwhile in an improved form - at the Department of Computer Science at the University of Karlsruhe since 1994. The course covers the whole design process from the behavioural level down to the layout synthesis. Its main subjects are programming of behavioural descriptions of circuits in VHDL and their validation by simulation.

The students usually have no problems to learn VHDL as it is similar to other imperative programming languages like Pascal or C. The concept of concurrent statements to describe parallelism, however, often causes disturbance. Also variables and signals are confused frequently.

Another interesting point is the programming style. Students of computer science seem to prefer a short, more algorithmic style as it is usual for computer programs. They usually do not realize the influence of the style on the resulting hardware effort. Students of electrical engineering, on the other hand, tend to start at the register-transfer level ignoring the capabilities of the system's high-level synthesis. Their programs are often more complex and difficult to understand.

Another problem the students had to contend with were test-benches for the simulation. Here, they often made mistakes and lost a lot of time. Hence, we decided to offer correct test-benches, which reduced also our effort to check the results. We also refrained from teaching advanced capabilities of Synopsys like the simulator's control language as the students usually failed to distinguish between the control language and VHDL. Instead, we wrote some routines which support the simulation, e.g. by additional control windows.

A crucial point to run a practical course successfully is a given, uniform environment for all tasks. Otherwise, the students may become totally confused by the nearly infinite possibilities of setup files, default values, and directory structures within a design system like Synopsys. This would also make troubleshooting impossible for us. Hence, we wrote an installation script that generates a consistent structure of directories, subdirectories, and setup files for each task.

REFERENCES

[1] Synopsys Inc.: http://www.synopsys.com/

[2] M.C. McFarland, A.C. Parker and R. Camposano, *The High-Level Synthesis of Digital Systems,* Proc. of the IEEE, Vol. 78, No. 2, Feb. 1990, pp. 301-318

[3] R.K. Brayton, G.D. Hachtel and A.L. Sangiovanni-Vincentelli, *Multilevel Logic Synthesis,* Proc. of the IEEE, Vol. 78, No. 2, Feb. 1990, pp. 264-300

[4] *IEEE Standard VHDL Language Reference Manual.* IEEE Std 1076-1987

[5] Cadence Inc.: http://www.cadence.com/

[6] Europractice: http://www.imec.be/europractice/europractice.html

[7] G. Lehmann, B. Wunder and M. Selz, *Schaltungsdesign mit VHDL: Synthese, Simulation und Dokumentation digitaler Schaltungen,* Poing: Franzis, 1994, P. 317 (in German)

[8] P. Kurup and T. Abbasi, *Logic Synthesis Using Synopsys,* Kluwer, 1995, P. 304

[9] Y.-C. Hsu, K.F. Tsai, J.T. Liu and E.S. Lin, *VHDL Modeling for Digital Design Synthesis,* Kluwer, 1995, P. 356

[10] A. Zell, *Simulation Neuronaler Netze,* Addison-Wesley, 1994, pp. 197-206 (in German)

[11] *IEEE Standard Test Access Port and Boundary-Scan Architecture*, IEEE Standard 1149.1-1990

Contribution:

Nacer Abouchi
CPE Lyon, LISA CNRS EP 0092
43 Bd. du 11 Nov. 1918
BP 2077 Villeurbqnne
FRANCE

M.A. Aguirre
Universidad de Sevilla
Avda. Reina Mercedes s/n
41012 Sevilla
SPAIN

Paul Amblard
Departement d'Informatique
Universite de Grenoble
LSR IMAG BP 53
F 38041 Grenoble Cedex 9
FRANCE

D. Auvergne
Laboratoire d'Informatique, de Robotique
et de Microélectronique de Montpellier
UMR 5506 UM2/CNRS, 161 rue ADA
34392 Montpellier Cedex 5
FRANCE

N. Azemard
Laboratoire d'Informatique, de Robotique
et de Microélectronique de Montpellier
UMR 5506 UM2/CNRS, 161 rue ADA
34392 Montpellier Cedex 5
FRANCE

Wojciech Białas
Faculty of Physics and Nuclear Techniques
University of Mining and Metallurgy
al. Mickiewicza 30
30-059 Kraków
POLAND

Mykola B. Blyzniuk
CAD Department
State University "Lviv Polytechnic"
12 Bandera Str., 290646 Lviv
UKRAINE

Anna Boszko
Institute of Electron Technology
Al. Lotników 32/46
PL-02-668 Warszawa
POLAND

S.A. Bota
EME-Departament
de Física Aplicada i Electrònica
Universitat de Barcelona
Av. Diagonal, 645-647. 08028 Barcelona
SPAIN

Marina Brik
Technical University of Tallinn
Ehitajate tee 5
EE0026, Tallinn
ESTONIA

Joan Cabestany
Universitat Politècnica de Catalunya
Gran Capitá s/n, Building C4
08034 Barcelona
SPAIN

J. Calderer
Departament d'Enginyeria Electrònica
Universitat Politècnica de Catalunya
C Gran Capitan s/n. 08034 Barcelona
SPAIN

P.F. Calmon
Atelier Interuniversitaire
de Micro-électronique de Toulouse
Campus INSA - Complexe Scientifique
de Rangueil - 31077 Toulouse
FRANCE

E. Cantó
Universitat Politècnica de Catalunya
Gran Capitá s/n, Building C4
08034 Barcelona
SPAIN

Jean-Pierre Chante
CPE Lyon, LISA CNRS EP 0092
43 Bd. du 11 Nov. 1918
BP 2077 Villeurbqnne
FRANCE

Gea-Ok Cho
ASIC Center
Corporate Technical Operations
Samsung Electronics CO., Ltd
KOREA

S. Cremoux
LIRMM : Laboratoire d'Informatique,
de Robotique et de Microélectronique
de Montpellier
UMR 5506 UM2/CNRS, 161 rue ADA
34392 Montpellier Cedex 5,
FRANCE

Dionizy Czekaj
University of Silesia
Institute of Engineering Problems
2, Śnieżna St.
Sosnowiec, PL 41-200,
POLAND

Adam Dąbrowski
Division for Signal Processing
and Electronic Systems
Poznań University of Technology
ul. Piotrowo 3a, 60-965 Poznań
POLAND

Władysław Dąbrowski
Faculty of Physics and Nuclear Techniques
University of Mining and Metallurgy
al. Mickiewicza 30, 30-059 Kraków
POLAND

Gilbert De Mey
University of Gent
ELIS, Sint Pieterniuwstraat 41
B-9000 Gent
BELGIUM

Rafał Długosz
Division for Signal Processing
and Electronic Systems
Poznań University of Technology
ul. Piotrowo 3a, 60-965 Poznań
POLAND

Julian Dudek
University of Silesia
Institute of Engineering Problems
2, Śnieżna St., Sosnowiec, PL 41-200,
POLAND

P. van Duong
MIKRON GmbH
Breslauer Strasse 1-3
85386 Eching
GERMANY

Kamran Eshraghian
Edith Cowan University
Joondalup WA 6027
AUSTRALIA

Daniel Esteve
Laboratoire d'Analyse et d'Architecture
des Systèmes du Centre National
de la Recherche Scientifique LAAS/CNRS
07, Avenue Colonel Roche
31077 Toulouse Cedex 4
FRANCE

J. Faura
SIDSA, Parque Tecnológico de Madrid
28760 Tres Cantos (Madrid)
SPAIN

A. Ferreira-Noullet
Institut des Sciences Appliquées de Toulouse
Département de Génie Electrique et Informatique
Campus INSA
Complexe Scientifique de Rangueil
31077 Toulouse
FRANCE

Jürgen Frickel
Institute for Computer Aided Circuit Design
University of Erlangen-Nuremberg
Cauerstrasse 6, 91058 Erlangen
GERMANY

Wolfram Glauert
Institute for Computer Aided Circuit Design
University of Erlangen-Nuremberg
Cauerstrasse 6, 91058 Erlangen
GERMANY

Spartacus Gomáriz
Department of Electronic Engineering
Technical University of Catalunya
Building C4, c/Gran Capità s/n
08034 - Barcelona
SPAIN

Aleksandr V. Gorish
Research Institute of Physical Measurements
of the Space Agency
Moscow
RUSSIA

Piotr Grad
University of Technology and Agriculture
Institute of Telecommunication
ul. Kaliskiego 7, 85-763 Bydgoszcz
POLAND

Mariusz Grecki
Technical University of Łódź
Department of Microelectronics
and Computer Science
Al. Politechniki 11, 93-590 Łódź
POLAND

Richard Grisel
CPE Lyon, LISA CNRS EP 0092
43 Bd. du 11 Nov. 1918
BP 2077 Villeurbqnne
FRANCE

Paweł Gryboś
Faculty of Physics and Nuclear Techniques
University of Mining and Metallurgy
al. Mickiewicza 30, 30-059 Kraków
POLAND

A. Herms
EME-Departament
de Física Aplicada i Electrònica
Universitat de Barcelona
Av. Diagonal, 645-647. 08028 Barcelona
SPAIN

R. Holgado
EME-Departament
de Física Aplicada i Electrònica
Universitat de Barcelona
Av. Diagonal, 645-647. 08028 Barcelona
SPAIN

Marek Idzik
Faculty of Physics and Nuclear Techniques
University of Mining and Metallurgy
al. Mickiewicza 30, 30-059 Kraków
POLAND

J.M. Insenser
SIDSA, Parque Tecnológico de Madrid
28760 Tres Cantos (Madrid)
SPAIN

Grzegorz Jabłoński
Technical University of Łódź
Department of Microelectronics
and Computer Science
al. Politechniki 11, 93-590 Łódź
POLAND

Jacek Jakusz
Faculty of Electronics Telecommunications
and Informatics
Technical University of Gdańsk
ul. G. Narutowicza 11/12, 80-952 Gdańsk
POLAND

Marcin Janicki
Technical University of Łódź
Department of Microelectronics
and Computer Science
al. Politechniki 11, 93-590 Łódź
POLAND

Gert Jervan
Technical University of Tallinn
Ehitajate tee 5, EE0026, Tallinn
ESTONIA

Igor Katchan
University of Karlsruhe
Institute of Computer Design
and Fault Tolerance
P.O. Box 6980, D-76128 Karlsruhe
GERMANY

Irena Y. Kazymyra
CAD Department
State University "Lviv Polytechnic"
12 Bandera Str., 290646 Lviv
UKRAINE

Chun-Sup Kim
ASIC Center
Corporate Technical Operations
Samsung Electronics CO., Ltd
KOREA

Yong-Hwan Kim
ASIC Center, Corporate Technical Operations
Samsung Electronics CO., Ltd
KOREA

Yuri N. Koptev
Research Institute of Physical Measurements
of the Space Agency
Moscow
RUSSIA

Vladimir A. Koval
CAD Department
State University "Lviv Polytechnic"
12 Bandera Str., 290646 Lviv
UKRAINE

Aleksandr A. Kuprienko
Rostov State University
Scientific-Design-Technological Department
"PIEZOPRIBOR"
5, Zorge St., Rostov on Don, SU 344104
RUSSIA

Stanisław Kuta
University of Mining and Metallurgy
Department of Electronics
30-059 Kraków, al. Mickiewicza 30
POLAND

Stefan W. Lachowicz
Edith Cowan University
Joondalup WA 6027
AUSTRALIA

Jose F. López
University of Las Palmas de Gran Canaria
35017-Las Palmas de Gran Canaria
SPAIN

Witold Machowski
University of Mining and Metallurgy
Department of Electronics
30-059 Kraków, al. Mickiewicza 30
POLAND

Jordi Madrenas
Department of Electronic Engineering
Technical University of Catalunya
Building C4, c/Gran Capità s/n
08034 - Barcelona
SPAIN

Jarosław Majewski
University of Technology and Agriculture
Institute of Telecommunication
ul. Kaliskiego 7, 85-763 Bydgoszcz
POLAND

Antti Markus,
Technical University of Tallinn
Ehitajate tee 5, EE0026, Tallinn
ESTONIA

Francesc Masana
GDS-DEE, UPC
C/J. Girona 1-3, C-4, 08034 Barcelona
SPAIN

Andrzej Materka
Institute of Electronics
Technical University of Lodz
Stefanowskiego 18, 90-537 Lodz
POLAND

Frank Mayer
University of Karlsruhe
Institute of Computer Design
and Fault Tolerance
P.O. Box 6980, D-76128 Karlsruhe
GERMANY

Jafaar Mejri
Institute for Computer Aided Circuit Design
University of Erlangen-Nuremberg
Cauerstrasse 6, 91058 Erlangen
GERMANY

Jaroslaw Mirkowski
Department of Computer Engineering
and Electronics
Technical University of Zielona Góra
ul. Podgorna 50, 65-246 Zielona Gora
POLAND

Juan Manuel Moreno
Universitat Politècnica de Catalunya
Gran Capitá s/n, Building C4
08034 Barcelona
SPAIN

Manuel Moreno
EME-Departament
de Física Aplicada i Electrònica
Universitat de Barcelona
Av. Diagonal, 645-647. 08028 Barcelona
SPAIN

Andrzej Napieralski
Technical University of Łódź
Department of Microelectronics
and Computer Science
al. Politechniki 11, 93-590 Łódź
POLAND

Jean-Louis Noullet
Laboratoire d'Analyse et d'Architecture
des Systèmes du CNRS
7, av. du Colonel Roche - 31077
Toulouse Cedex 4
FRANCE

Ole Olesen
Technical University of Denmark
Center for Integrated Electronics
Building 344, DK - 2800 Lyngby
DENMARK

Leszek J. Opalski
Institute of Electronic Fundamentals
Faculty of Electronics
and Information Technology
Warsaw University of Technology
ul. Nowowiejska 15/19, 00-665 Warszawa
POLAND

Mariusz Orlikowski
Technical University of Łódź
Department of Microelectronics
and Computer Science
al. Politechniki 11, 93-590 Łódź
POLAND

M. Padeffke
Friedrich Alexander University
of Erlangen-Nuremberg
Institute for Computer Aided Circuit Design
Cauerstrasse 6, 91058 Erlangen
GERMANY

Anatoli E. Panich
Rostov State University
Scientific-Design-Technological Department
"PIEZOPRIBOR"
5, Zorge St., Rostov on Don, SU 344104
RUSSIA

Bogdan Pankiewicz
Faculty of Electronics Telecommunications
and Informatics
Technical University of Gdańsk
ul. G. Narutowicza 11/12, 80-952 Gdańsk
POLAND

Pawel Pełczynski
Institute of Electronics
Technical University of Lodz
Stefanowskiego 18, 90-537 Lodz
POLAND

Hans Jörg Pfleiderer
University of Ulm
D-89069 Ulm
GERMANY

Christophe Premont
CPE Lyon, LISA CNRS EP 0092
43 Bd. du 11 Nov. 1918
BP 2077 Villeurbqnne
FFANCE

Jaan Raik
Technical University of Tallinn
Ehitajate tee 5, EE0026, Tallinn
ESTONIA

Abdul Wahab A. Salman
Technical University of Łódź
Department of Microelectronics
and Computer Science
al. Politechniki 11, 93-590 Łódź
POLAND

Roberto Sarmiento
University of Las Palmas de Gran Canaria
35017-Las Palmas de Gran Canaria
SPAIN

Detlef Schmid
University of Karlsruhe
Institute of Computer Design
and Fault Tolerance
P.O. Box 6980, D-76128 Karlsruhe
GERMANY

Noureddine Senouci
Laboratoire d'Analyse et d'Architecture
des Systèmes du Centre National
de la Recherche Scientifique LAAS/CNRS
07, Avenue Colonel Roche
31077 Toulouse Cedex4
FRANCE

Zbigniew Skowronski
Department of Computer Engineering
and Electronics
Technical University of Zielona Góra
ul. Podgorna 50, 65-246 Zielona Gora
POLAND

Bang-Sup Song
University of Illinois,
Urbana, IL
USA

Michał Strzelecki
Institute of Electronics
Technical University of Lodz
Stefanowskiego 18, 90-537 Lodz
POLAND

Zygmunt Surowiak
University of Silesia
Institute of Engineering Problems
2, Śnieżna St., Sosnowiec, PL 41-200,
POLAND

Stanisław Szczepański
Faculty of Electronics Telecommunications
and Informatics
Technical University of Gdańsk
ul. G. Narutowicza 11/12, 80-952 Gdańsk
POLAND

Francis Therez
Laboratoire d'Analyse et d'Architecture
des Systèmes du Centre National
de la Recherche Scientifique LAAS/CNRS
07, Avenue Colonel Roche
31077 Toulouse Cedex4
FRANCE

Eric Tournier
Laboratoire d'Analyse
et d'Architecture des Systčmes du CNRS
7, av. du Colonel Roche
31077 Toulouse Cedex 4
FRANCE

Marek Turowski
Technical University of Łódź
Department of Microelectronics
and Computer Science
al. Politechniki 11, 93-590 Łódź
POLAND

Raimund Ubar
Technical University of Tallinn
Ehitajate tee 5, EE0026, Tallinn
ESTONIA

Bogusław Więcek
Technical University of Łódź
Institute of Electronics
ul. Stefanowskiego 18/22, 90-924
POLAND

Ryszard Wojtyna
University of Technology and Agriculture
Institute of Telecommunication
ul. Kaliskiego 7, 85-763 Bydgoszcz
POLAND

Wojciech Wójciak
Technical University of Łód
Department of Microelectronics
and Computer Science
al. Politechniki 11, 93-590 Łódź
POLAND

Robert Wydmański
University of Mining and Metallurgy
Department of Electronics
30-059 Kraków, al. Mickiewicza 30
POLAND

Mariusz Zubert
Technical University of Łódź
Department of Microelectronics
and Computer Science
al. Politechniki 11, 93-590 Łódź
POLAND

Index:

GPSR Compliance

The European Union's (EU) General Product Safety Regulation (GPSR) is a set of rules that requires consumer products to be safe and our obligations to ensure this.

If you have any concerns about our products, you can contact us on ProductSafety@springernature.com

In case Publisher is established outside the EU, the EU authorized representative is:

Springer Nature Customer Service Center GmbH
Europaplatz 3
69115 Heidelberg, Germany

Batch number: 10704713

Printed by Printforce, the Netherlands